MATHEMATICS REBOOTED

MATHEMATICS REBOOTED

A Fresh Approach to Understanding

LARA ALCOCK

OXFORD
UNIVERSITY PRESS

OXFORD
UNIVERSITY PRESS

Great Clarendon Street, Oxford, OX2 6DP,
United Kingdom

Oxford University Press is a department of the University of Oxford.
It furthers the University's objective of excellence in research, scholarship,
and education by publishing worldwide. Oxford is a registered trade mark of
Oxford University Press in the UK and in certain other countries

First Edition published in 2017

Impression: 1

Published in the United States of America by Oxford University Press
198 Madison Avenue, New York, NY 10016, United States of America

British Library Cataloguing in Publication Data
Data available

Library of Congress Control Number: 2017934732

ISBN 978-0-19-880379-9

Printed and bound by
CPI Group (UK) Ltd, Croydon, CR0 4YY

PREFACE

What do adults choose to learn? Interests are personal, but many learn languages, or read about science, history, politics, economics, philosophy, or psychology. And many enjoy both classic and contemporary art, literature, plays, and films. Not everyone values these things equally, but many are proud to know at least a little about each. I've written this book because I would like to see mathematics added to the list. Not because mathematics is a standard school subject, but because it is a cumulative intellectual endeavour with a long history and a wealth of clever and interesting ideas. I wouldn't expect a layperson to know many details or to have any idea about the cutting edge, but I'd like more people to feel that they could speak confidently about key mathematical concepts and approaches to reasoning.

I got into a position to write this book by studying mathematics then mathematics education. Mathematics education as an academic discipline overlaps with mathematical cognition as a branch of psychology: both study ways in which people learn and think about mathematics. And research in these areas has revealed a lot. We know quite a bit about typical errors, misconceptions, and sources of confusion in children and in adults. We have good theories about how some of these arise, and we're testing them with intervention studies designed to improve teaching and learning. Education is complicated, of course—anyone who has been in a class of 30 teenagers knows that intellectual development requires more than good lesson plans. But teachers and researchers know about numerous stumbling blocks in mathematical thinking, and they will recognize much of the content in this book.

That said, this is not a book about research—I use my knowledge about both mathematics and education in a more cavalier way than I would in academic writing. I explain why some ideas are naturally confusing, but this book is essentially an account of how I think about the subject.

Like every teacher, my thinking is heavily influenced by my early experiences, and I do not try to hide that—I point out places in which I suspect that my way of understanding is idiosyncratic. But I don't include jokes, puns, or attempts to make mathematics interesting. In my view, there is no need to make mathematics interesting—it is fascinating all by itself.

This book has multiple intended audiences, so readers with mathematical backgrounds will notice that sometimes, when I introduce an idea, I skate over the subtleties. That's deliberate: I think it can be important to consolidate a simple version first. Sometimes the subtleties don't appear until considerably later in the book, so I hope that such readers will be patient. In particular, people who have studied higher level mathematics will have been told to be wary of intuition based on visual representations. That is sensible advice, and mathematicians offer it when they want students to question their assumptions and to justify their ideas within an established theory. Teaching disciplined reasoning within established theories is a valid aim, but it's not mine in this book. My aim here is to communicate with nonspecialists about mathematical ideas. And I really like pictures, so I use them a lot, albeit discussing their limitations. Similarly, I start each main chapter with basic ideas, but I mean basic in an everyday rather than mathematically foundational sense. In my experience, learners need to work down to foundational ideas just as they need to work up to advanced ones. So the more foundational discussions appear at the end of the book, not the beginning.

To conclude this Preface, I would like to thank many friends and colleagues for their help and feedback. For carrying this book through the practicalities from proposal to finished product, thank you to Dan Taber of Oxford University Press. For the usual extraordinarily patient and attentive copyediting and typesetting, thank you to Charles Lauder Jr., and to Karen Moore and her team. For extremely valuable feedback on drafts of the content, thank you to the reviewers of the original proposal, and to Nina Attridge, Sophie Batchelor, Louisa Butt, Jane Coleman, Lucy Cragg, Jo Eaves, Ant Edwards, Cameron Howat, Hazel Howat, Matthew Inglis, Jayne Pickering, Artie Prendergast-Smith, and David Sirl. Their input taught me that educated nonspecialists find algebra and logic easier than

I expected, but diagrams harder. This, together with numerous detailed comments, improved my writing in ways that I hope will help all readers.

Finally, I'd like to dedicate this book to my parents, Angela and Eric Alcock, who always supported me but never pushed. That allowed me to develop a genuine love for mathematics, and I'm grateful for it.

CONTENTS

Introduction   *xiii*

1 Multiplying   **1**
   1.1 Famous theorems   1
   1.2 Multiplication made easy   2
   1.3 Properties of multiplication   5
   1.4 'Multiplication makes things bigger'   9
   1.5 Squares   15
   1.6 Triangles   21
   1.7 Pythagoras' theorem   26
   1.8 Pythagorean triples   31
   1.9 Fermat's Last Theorem   36
  1.10 Review   39

2 Shapes   **41**
   2.1 Tessellations   41
   2.2 Regular polygons   43
   2.3 Regular tessellations   47
   2.4 Interior angles   51
   2.5 Mathematical theory building   56
   2.6 Semi-regular tessellations   57
   2.7 More semi-regular tessellations   62
   2.8 Algebra and rounding   66
   2.9 Symmetry: Translations and rotations   68
  2.10 Symmetry: Reflections and groups   73
  2.11 Symmetry in other contexts   77
  2.12 Review   79

3 Adding up 81

   3.1 Infinite sums 81

   3.2 Fractions 83

   3.3 Adding fractions 88

   3.4 Adding up lots of numbers 93

   3.5 Adding up lots of odd numbers 97

   3.6 Powers of 2 101

   3.7 Adding up powers 105

   3.8 The geometric series $1+\frac{1}{2}+\frac{1}{4}+\frac{1}{8}+\frac{1}{16}+\frac{1}{32}+\ldots$ 107

   3.9 The harmonic series $1+\frac{1}{2}+\frac{1}{3}+\frac{1}{4}+\frac{1}{5}+\frac{1}{6}+\ldots$ 110

  3.10 Convergence and divergence 113

  3.11 Review 118

4 Graphs 121

   4.1 Optimization 121

   4.2 Plotting points 122

   4.3 Plotting graphs 126

   4.4 $y = mx + c$ (or b) 130

   4.5 More or less? 133

   4.6 Intersecting lines 137

   4.7 Areas and perimeters 141

   4.8 Area formulas and graphs 144

   4.9 Circles 147

  4.10 Polar coordinates 151

  4.11 Coordinates in three dimensions 154

  4.12 Review 157

5 Dividing 159

   5.1 Number systems 159

   5.2 Dividing by 9 in base 10 160

   5.3 If and only if 165

   5.4 Division and decimals 167

   5.5 Decimals and rational numbers 173

   5.6 Lowest terms 178

   5.7 Irrational numbers 183

   5.8 How many rationals and irrationals? 187

   5.9 Number systems 194

  5.10 Review 198

Conclusion   **201**
   Why didn't my teachers explain it like that?   202
   What is it all for?   203
   What do mathematicians do?   205
   What shall I read next?   207

References   *209*
Index   *221*

INTRODUCTION

Mathematics—ordinary, school-level mathematics—is elegant and interesting. Many people suspect this but were not able to appreciate it at the time. Some were put off by repetitive exercises. Others missed a key idea and found later work meaningless. Most, however, did learn something, and this book aims to build on that. It revisits basic ideas, avoiding repetition and focusing on meaning. It explores these ideas in depth, uses them to explain what mathematicians care about and how they think, and builds gradual links to more advanced material. In this brief Introduction I explain how this might work for people with different backgrounds, and what you should therefore expect.

Is this book for me?

This book is for several sets of people. It is for you if you liked mathematics up to a point but later lost track of it because you didn't understand a key idea or you didn't like your teacher or you'd just discovered the attractions of life outside school. It is for you if you liked mathematics but didn't pursue it beyond compulsory education because it wasn't your main interest—perhaps your thing was music or languages or psychology or design. Finally, it is for you if you liked mathematics and did pursue it, perhaps even to undergraduate level, but you've now forgotten much of what you once knew. Whether you find it an easy read will depend upon which group you fall into, but I'm pretty sure that everyone will get further than they'd think. I asked nonexperts for feedback on all of the chapters, and those who gave up mathematics at age 16 told me that they did struggle a bit, but only in the later sections, and that by then it didn't matter because they'd built up their confidence and were feeling good about themselves. Also, this Introduction will offer advice about reading mathematics effectively.

First, though, some comments on how the book might interest specific readers. It might help teachers who are not mathematics specialists but who would like to feel more confident about relationships between basic mathematics and more advanced concepts. It might help adults who are learning mathematics for qualifications, though for that audience it will do only part of the job. Its focus on mathematical ideas and their inter-relationships will, I hope, make anyone's learning easier and more interesting. But people studying for qualifications will also need to become fluent in certain types of calculation, and they will need another resource for that.

It might be useful for parents and carers who want to help their children understand mathematics. I hope that it is, but I would counsel caution and patience. If this book allows you to see mathematical links that you never noticed before, and at some point you experience an epiphany, that's great—it's exactly what I'm aiming for. But epiphanies are personal—they depend on the structure of an individual's knowledge. Someone who knows less than you do might not see the value of an insight. He or she might need to acquire new knowledge by increments and, depending on the order in which this is done, might eventually consider the key link unremarkable. I would like adults to talk with children about mathematics; I would not like children to be put off the subject by demands for enthusiasm. Personally, I enjoyed mathematics because people gave me interesting things to think about, then left me to it. I knew where to go when I had questions or something to say, but there was no immediate pressure to resolve everything. That's the approach I'd advocate, especially if you have a smart kid.

Finally, this book might interest precisely those smart kids—young people who enjoy mathematics and think that they might want to study it at higher levels. If that describes you, I'm glad you're here.

What if I'm not very good at mathematics?

I'm aware that some readers will be nervous. Perhaps you reached a point at which mathematics made you feel inadequate, and you have no desire to revisit that feeling. Perhaps you believe that you lack mathematical talent and are therefore doomed to failure. Perhaps you were given this book by well-intentioned friends and have only begun reading it so as not to appear rude. Whatever the case, I hope to convince you to read on, because mathematical thinking is not magical. It is often thought of that way in our culture, where it is common to have a demanding career or to run a happy and successful household, yet to say, 'Oh, I am terrible at maths.' I hear this a lot, and every time it is clear to me that it cannot really be true: this person is obviously a capable thinker. So, while I believe that many people didn't master all the mathematics they encountered in school, I also believe that the majority can understand more, and that, when learning without the pressure of exercises and tests, they are likely to enjoy it.

By way of analogy, I am terrible at tennis, and at any sport that involves hitting an object with another object. Some people are not like this—they can hit a tennis ball with power and control. I find this both admirable and utterly mysterious, and no amount of practice would make me that good. But that does not mean that I could not improve. I could get better at tennis—probably a lot better, especially if I had expert guidance. I expect that you agree about that, but people routinely say 'I can't' in the face of mathematics. I think this is partly because self-deprecation is expected: it doesn't do to show off. But I think that we do ourselves a disservice if we are swayed by an 'I can't' attitude, because there is no real reason to think of mathematics as qualitatively different from other endeavours. Indeed, the great thing about proper mathematical understanding is that it requires a lot *less* work than learning a physical skill. Although many people experience mathematics as a bunch of procedures to be learned through repetition, experts do not see it that way. They see well-defined concepts that fit together with a logic and elegance that makes mathematics inherently memorable. I hope to convey some of that logic, elegance and memorability in this book.

What should I expect?

This book explains and links mathematical ideas in an order that differs from a typical school curriculum. School mathematics tends to come in horizontal slices: children learn basic ideas about several topics, then, the next year, they learn slightly more advanced ideas about those topics, and so on.

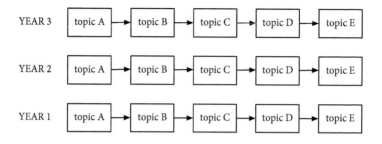

This is entirely sensible. But it means that the vertical links are not very salient, which is important because mathematics can be seen as a highly interconnected network in which more sophisticated ideas build upon more basic ones. So this book's approach is to focus explicitly on the vertical links. Each chapter starts with an idea that is bang in the middle of school mathematics—primary school mathematics in many cases—then takes a tour upward through related concepts, arriving eventually at ideas that people encounter in more advanced study (see the next diagram).

If that endpoint sounds daunting, don't worry—each chapter builds up gradually, drawing on research in education and psychology to explain common causes of confusion, and introducing representations that clarify what is going on and facilitate valid reasoning. I explain ideas where they are first introduced, and there is an extensive index so that you can locate such information if you forget it later. Also, the text contains numerous questions that you could follow up to explore the ideas in more depth. If that sounds good, maybe read with a pencil and paper to hand. If it doesn't, that's fine—I intend this to be a book that you can read on a sun-lounger or the train to work, so do ponder or read on as you like.

Also, it's not like you have to read everything. If you lost track of mathematics at some point, you might want to read just the first half of each chapter. It's fine to enjoy what is digestible and let it settle, then decide whether to have another go. If you gave up mathematics to pursue a different passion, you might enjoy the later sections in each chapter because they introduce ideas that you haven't seen before. If you studied mathematics to a higher level, there might be nothing truly new. But the daily grind of a pressured, test-focused environment can easily swamp the most beautiful ideas. In a book like this, I have the luxury of drawing

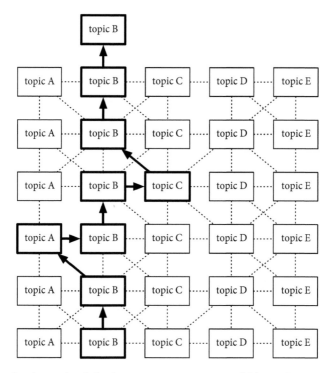

out the things that I think are most interesting, and I hope that you too will gain some new insight.

And that's why I'm writing—because I love mathematics and I want as many people as possible to see why. To achieve that, though, I'd like you to understand something about mathematical reading.

How can I read mathematics effectively?

Good mathematical reading does not look like ordinary reading. This becomes obvious when inspecting *scanpaths* that trace human eye movements. Below is the scanpath generated by a mathematician reading the instructions for a research study. This reading is fairly linear: the person reads a line from left to right, then moves to the next line and reads that, and so on.

During the first part of the experiment you will be asked to read a series of mathematical proofs, each written by a student in an examination.

Please read each proof and decide whether or not it is valid. When you are happy with your decision click the mouse button.

You should spend as long as you need reading each proof. Do not rush!

If you would like to speak as you read the proofs please feel free to do so.

If you get completely stuck, then click the mouse button to move on.

The first proof is for practice.

Click the mouse when you are ready to start.

Expert mathematical reading is not like this. Below is a scanpath[1] for the same mathematician reading a piece of mathematical text. Look at all those extra movements back and forth. These do not reflect poor reading ability—for this person, the text was pretty straightforward. Rather, they reflect an expert's search for logical relationships among all the concepts and claims. When reading about mathematics, everyone should feel the need to read back and forth considerably more than when reading ordinary text. This does not indicate mathematical incompetence—on the

[1] Both scanpaths were generated by the *eye tracker* in the Mathematics Education Centre at Loughborough University, where I work. An eye tracker is like a normal computer except that the screen records where the viewer is looking. It monitors this using infrared cameras, which is completely noninvasive—the viewer can't tell that it's happening (of course, we do tell research participants what we're up to). The image has dots and lines because vision is not 'smooth' in the way it appears to subjective experience—when a person reads static images, their eyes perform short stops called *fixations* and shift between these in rapid movements called *saccades*.

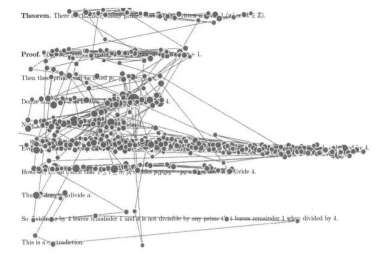

Theorem. There are infinitely many primes ... written as $\frac{a}{b}$ ... (x ... or $\wedge$ ... $\in \mathbb{Z}$).

Proof. ... $+ 1$.

Then this prime can be listed $p_1, ..., p_n$.

Define a ... as follows.

No ... under ...

Ev ... divisible by 4.

However, if ... such that $1 \le i \le n$, p_i divides $p_1 p_2 p_3 ... p_n$, like a, it divide 4.

Thus ... does ... divide a.

So a divided by 4 leaves remainder 1 and a is not divisible by any prime t ... t leaves remainder 1 when divided by 4.

This is a contradiction.

---

contrary, it indicates a mature search for understanding.[2] So, while you read this book, pause whenever you feel the need, read each line several times if that helps, review whole sections when that seems useful, and think about how you would explain it all to someone else. If you do that, your reading will be a bit slower but your understanding will grow a lot faster.

---

[2] Our research team is pretty sure about this because undergraduate mathematics students read in a similar way, but less so—they are not yet as careful as experts. When we gave undergraduates research-based *self-explanation training*, their eye movements became more expert-like, and they generated better explanations and performed better on a comprehension test. It was important that they deliberately think about the links across the mathematical text. If this research interests you, you can read about it in the *Notices of the American Mathematical Society* at doi.org/10.1090/noti1263, and there's more information at http://www.lboro.ac.uk/departments/mec/research/.

CHAPTER 1

Multiplying

1.1 Famous theorems

Have you heard of Pythagoras' theorem? How about Fermat's Last Theorem? Many people have, because these theorems appear frequently in popular accounts of mathematics—they have captivated mathematicians for centuries and are considered accessible for lay audiences. That does not mean, however, that everyone knows what they say or understands why mathematicians are so keen on them.

In this chapter I'll explain both the subtleties of these theorems and the reasons why they are considered impressive. Some readers might be sceptical about whether I'll manage that, and some might be right to feel that way—anyone who skipped the Introduction[1] will have missed my saying that I expect different people to read different amounts of each chapter. Maybe you'll get halfway through and decide that you've had enough. That's fine, but many people will get further than they'd think. Your mathematical knowledge might be rusty and full of holes, but people who can function well in our complicated world must be good general thinkers, and mathematics is just general thinking about abstract concepts.

At this stage, though, many readers won't know what a theorem is, and that's easy to fix: a theorem is a true mathematical statement, often one that says something about every case of a certain type. Here are some examples.

---

[1] I'm not offended if you did—I often skip introductions myself.

Theorem: The sum of two even numbers is always even.

Theorem: A number is divisible by 9 if and only if the sum of its digits is divisible by 9.

Did you know these things already? Do you know why they are true? We'll revisit the second in Chapter 5, examining the reasons carefully. For now I just want to make clear that theorems do not have to be about esoteric concepts that no one understands. Some are easy to understand, even if it is not easy to see why they are true. Both Pythagoras' Theorem and Fermat's Last Theorem are harder, but this chapter will work up to them, taking in numerous elementary ideas. We'll start right at the beginning, with straightforward multiplication.

1.2 Multiplication made easy

When I was little I had a Ladybird book called *Multiplication Made Easy*. I loved that book. It wasn't remarkable or complicated—all it did was provide pictures of objects laid out in arrays so that you could see and count, for instance, four rows of six things (six buttons, or six apples or feathers or stars).

What I liked about this was not the counting or even the pleasing arrangements of objects. What I liked was that by turning the page around, *you could instead see six rows of four*. And the same thing worked for five rows of seven (seven rows of five), and for two rows of ten (ten rows of two), and for every pair of numbers in the book. This, I thought, was *brilliant*. At the time I couldn't have told you why, but I still think it's brilliant, and now I can: the swapping relationship is tremendously labour-saving, and the array representation provides insight about why it works.

The saving arises because the fact that 4×6 is the same as 6×4 (and so on) means that only 55 of the first 100 multiplication facts are really different. The memory work is not quite halved because swapping buys us nothing for calculations like 7×7, but still that's an impressive reduction.

×	1	2	3	4	5	6	7	8	9	10
1	1	2	3	4	5	6	7	8	9	10
2	2	4	6	8	10	12	14	16	18	20
3	3	6	9	12	15	18	21	24	27	30
4	4	8	12	16	20	24	28	32	36	40
5	5	10	15	20	25	30	35	40	45	50
6	6	12	18	24	30	36	42	48	54	60
7	7	14	21	28	35	42	49	56	63	70
8	8	16	24	32	40	48	56	64	72	80
9	9	18	27	36	45	54	63	72	81	90
10	10	20	30	40	50	60	70	80	90	100

Indeed, it is possible to do better. The 1s are straightforward, and the 10s probably don't need memorizing either. So we're down to 36 facts, which is far fewer than 100.

×	1	2	3	4	5	6	7	8	9	10
1	1	2	3	4	5	6	7	8	9	10
2	2	4	6	8	10	12	14	16	18	20
3	3	6	9	12	15	18	21	24	27	30
4	4	8	12	16	20	24	28	32	36	40
5	5	10	15	20	25	30	35	40	45	50
6	6	12	18	24	30	36	42	48	54	60
7	7	14	21	28	35	42	49	56	63	70
8	8	16	24	32	40	48	56	64	72	80
9	9	18	27	36	45	54	63	72	81	90
10	10	20	30	40	50	60	70	80	90	100

This appeals to my lazy side. And laziness, I think, is a characteristic shared by many mathematicians, who do not like to work harder than necessary. In particular, they do not like to memorize factual information—they prefer to reconstruct it using general relationships.

Such general relationships form one main theme of this book, which will highlight ways in which they can be captured by good representations. In my opinion, the dot array is a good representation for at least three reasons. First, in an array showing four rows of six objects, the four sixes are clearly visible. That might not seem remarkable, but it is: try seeing four sixes in a single line of dots,

or in a random jumble of them.

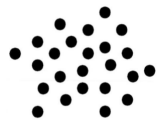

It can be done, but it's much harder.

Second, and more importantly, the array shows not only four sixes, but also six fours. Seeing both requires the right kind of looking, but this can be facilitated by turning the page around or by bracketing the dots in different ways. Turning or bracketing does not change the number of dots; a mathematician would say that the number is *invariant* under this change.

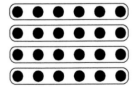

Third, the array can be treated as *generic* in the sense that there is nothing special about the 4 and the 6—an array will have the same properties for other pairs of numbers. Without needing dots on a page, I am confident that 5×3 must equal 3×5, and that 7×129 must equal 129×7. I can't 'see' a row of 129 dots, and I neither know nor care what 7×129 is, but I can imagine the array and I'm confident that whatever it is, it's equal to 129×7. For me, the array makes the swapping relationship obvious and therefore easy to remember.

I do not claim, of course, that this or any other representation has the same effect for everyone. A representation on a page is inert. The dots just sit there, and the action takes place in the viewer's mind—you have to look at the diagram in a certain way. Some representations in this book are more abstract and require more effortful thought. But I will argue that it's worth learning to look in the right way because doing so can reduce memory load and provide a big payoff in insight.

1.3 Properties of multiplication

As you will have gathered, I think that the 'answer' to 7×129 is less interesting than the fact that for any two numbers n and m, it is true that $n \times m = m \times n$. This sort of higher-level regularity appeals to mathematicians, and the swapping property has a mathematical name: *commutativity* ('com-mute-a-tivity'). Mathematicians say 'multiplication is commutative', and mathematics students learn words like 'commutative' when they arrive at university. Well, they should, but many don't, at least for a while—you don't need to know this fancy name in order to use the property in calculations or problem solving.

But naming is useful because it highlights commonalities across concepts. For instance, addition is commutative too: for any two numbers n and m, it is true that $n+m = m+n$. Addition doesn't lend itself to an array representation, but we can still illustrate this by arranging and bracketing dots. Here is a diagram for $7 + 3 = 3 + 7$. To see $3 + 7$, just turn the page upside down.

Again the number of dots is invariant, so the resulting sum must be the same. And again the diagram can be seen as generic. Thinking about similar diagrams convinces me that 127492 + 3996 must equal 3996 + 127492, whatever that is.

So multiplication and addition are both commutative. In fact, they are both *commutative binary operations*. They are called *binary operations* because they each take two numbers (hence 'binary') and operate with them to give another (hence 'operation'). Subtraction is a binary operation too, but it is *not* true that 7 − 3 equals 3 − 7 or, in general, that $n - m = m - n$. Subtraction is not commutative.

Here we will stick with multiplication and addition, which are linked by a further general property called *distributivity* ('distrib-ute-ivity'). Mathematicians say that 'multiplication distributes over addition', meaning that, for instance,

$$4 \times (3 + 5) = (4 \times 3) + (4 \times 5).$$

I find it helpful to 'read' the brackets[2] in expressions like this by slowing down in some places and speeding up in others, saying[3]

'four times three-plus-five equals
four-times-three plus four-times-five.'

Distributivity, like commutativity, can be represented with an appropriate array.

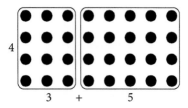

---

[2] British people tend to say 'brackets' regardless of their shapes: ( ) or [ ] or { }. North Americans tend to say 'parentheses' for the round ones, but I'm British so I'll stick with what I'm used to.

[3] When I do this in lectures, students laugh at me but then start doing it themselves.

Those who remember working on brackets will also know that $4 \times (3+5)$ is often written as $4(3+5)$ (juxtaposition denotes multiplication). And, by convention, multiplication takes priority over addition, so $(4 \times 3)+(4 \times 5)$ could be written without brackets as $4 \times 3 + 4 \times 5$. Using brackets might therefore make you uncomfortable—you might feel that $4 \times 3 + 4 \times 5$ is 'better' because it uses fewer symbols and because a mathematician would know what it meant, meaning that $(4 \times 3) + (4 \times 5)$ is inferior or less grown up.

I'd encourage common sense in such matters, though. Some notations are good for some things, and some are good for others. Some are very brief, but their brevity obscures key structures. I think that this is true of $4 \times 3 + 4 \times 5$, which is horrible on the eye because the symbols '$\times$' and '$+$' look similar and because English is read from left to right— both of these things invite interpretation errors. For me, the brackets in $(4 \times 3) + (4 \times 5)$, while not mathematically necessary, make it easier to focus on the meaning of the expression. Similarly, if you prefer $4 \times (3 + 5)$ to $4(3 + 5)$, by all means use it. It's useful to learn notational conventions; it's also useful to know that mathematicians, while respecting correctness, select representations depending on what they want to see or achieve.

With that in mind, distributivity can be extended further. For instance,

$$(4 + 2) \times (3 + 5) = (4 \times 3) + (4 \times 5) + (2 \times 3) + (2 \times 5).$$

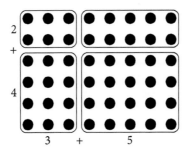

At this point the dots get a bit cumbersome, and it helps to move to an area representation with squares. This doesn't change anything, really,[4] but it generalizes better. Here is an area diagram for $4 \times (3 + 5)$.

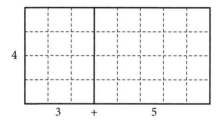

And here is one for $(4 + 2) \times (3 + 5)$.

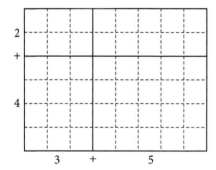

Again, using areas to think about multiplication might seem counter to what you learned in school. Probably you learned first to multiply, then to use multiplication to calculate areas. But this link can be reversed to gain insights about multiplication. To facilitate that, I will label the rows and columns as axes are labelled in graphs, with (0, 0) in the bottom left corner (more on graphs in Chapter 4). You can imagine starting at 0 and moving to the right or up, counting at every full square. This will help when thinking further about properties of multiplication.

---

[4] I exaggerate—it changes something rather subtle, but I'll discuss that later.

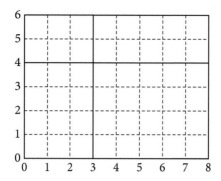

1.4 'Multiplication makes things bigger'

You probably learned as a child that multiplication makes things bigger. Hopefully no one in authority actually said this, but you will have learned it anyway, just by exposure. For most early multiplication tasks, the answer is bigger than both original numbers, sometimes by a lot.

To say 'multiplication makes things bigger' is, of course, inaccurate. Multiplication doesn't change the original numbers, it just combines them to give another (it's a binary operation). A person who says 'multiplication makes things bigger' really means 'the result of multiplying together two numbers is another number that's bigger than either of those you started with.' That's a mouthful, though, so we can probably be charitable about the abbreviation. Unfortunately, the claim is not only inaccurate but also untrue. It doesn't hold even for the tasks given to children. For instance, $7 \times 1 = 7$, and in general multiplying by 1 does not 'change' a number at all. But people tend to ignore such cases—which are sometimes called *degenerate*—because everyone feels that multiplying by 1 isn't really doing anything.

So the idea that multiplication makes things bigger is reasonable in early arithmetic. In that restricted context, it is a useful *heuristic*. And such heuristics are learned naturally. Human beings take numerous varied experiences and notice what is common among them: we infer generalizations that will render life predictable. We can't switch off this tendency and it would be mad to want to. But the fact that it works well and largely unconsciously means that we are vulnerable to learning 'facts' that aren't really true. If an early pattern does not extend as we expect, the heuristic can become a *cognitive obstacle* that interferes with later thinking. The idea that 'multiplication makes things bigger' is one such heuristic, because it does not hold for numbers that are less than 1.

Consider, for instance, $\frac{1}{2} \times 6$. To represent this using area requires a rectangle with one side of length $\frac{1}{2}$ and one of length 6, as shown in the following image. I've made the squares bigger so that the number $\frac{1}{2}$ fits between 0 and 1, and the result is the shaded area comprising six half-squares. Rearranged, these would give three whole squares, so $\frac{1}{2} \times 6 = 3$. In relation to the 'multiplication makes things bigger' heuristic, this is a bit of a mess: the result is bigger than $\frac{1}{2}$ but smaller than 6.

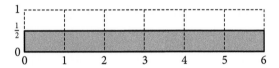

Note, though, that the other properties discussed earlier do *not* change. Commutativity and distributivity are more than useful heuristics; they are fundamental properties of multiplication and addition. This is illustrated in the following diagrams. Can you see these diagrams as generic? How would the same things work for other numbers?

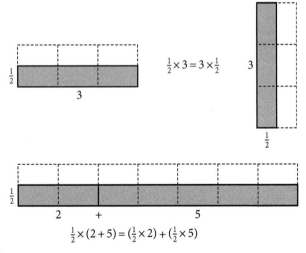

$$\frac{1}{2} \times 3 = 3 \times \frac{1}{2}$$

$$\frac{1}{2} \times (2 + 5) = (\frac{1}{2} \times 2) + (\frac{1}{2} \times 5)$$

Note also that the area representation holds up well—it is useful for multiplying whole numbers by fractions. And it comes into its own for multiplying fractions by fractions. The obvious place to start is with $\frac{1}{2} \times \frac{1}{2}$, but people routinely get confused by that because it evokes two unhelpful intuitions. The first is the faulty heuristic that multiplication makes things bigger. The second is the feeling that when two halves are kicking around, the 'answer' should be 1. But that is not right either. Certainly $\frac{1}{2} + \frac{1}{2}$ is equal to 1, but that tells us nothing about $\frac{1}{2} \times \frac{1}{2}$.

In actual fact, $\frac{1}{2} \times \frac{1}{2} = \frac{1}{4}$. I don't recall when I first learned this, but I know that I wasn't happy about it. It felt wrong, probably for the reasons I just stated. I didn't give up easily, though, and at some point I sorted it out with alternative, more simplistic, phrasing. When I first encountered multiplication, no one said 'multiplied by' or even 'times'. Instead they said 'lots of', where 'three lots of five' meant 3×5. I found it helpful to apply this to $\frac{1}{2} \times \frac{1}{2}$. If I said 'a half *lots of* a half', that was very like saying 'a half *of* a half', which did seem to be a quarter.

Now, though, I'd think about $\frac{1}{2} \times \frac{1}{2}$ using area. In the following diagram, the area of the big square is $1 \times 1 = 1$. The area of the smaller shaded square is $\frac{1}{2} \times \frac{1}{2}$. Exactly four copies of it fit into the big square, so the shaded area must be $\frac{1}{4}$. If you never quite understood why $\frac{1}{2} \times \frac{1}{2} = \frac{1}{4}$, I hope that sorts it out for you.

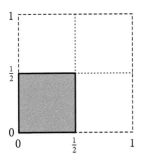

We're not stopping there, though, because there's more mileage in this way of thinking about fraction multiplication. Suppose we're interested in $\frac{2}{5} \times \frac{3}{4}$, for which an area diagram appears below. The left vertical side is divided into five equal parts (fifths), so we can set up one side of length $\frac{2}{5}$. The bottom is divided into four equal parts (quarters), so we can set up another side of length $\frac{3}{4}$. Again, the result is shaded. What fraction of the 1×1 square is that?

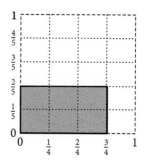

Dividing one side of the square into five equal parts and the other into four gives $5 \times 4 = 20$ equal parts in total, each of which must have area $\frac{1}{20}$. The multiplying results in $2 \times 3 = 6$ of the 20 equal parts being shaded. So $\frac{2}{5} \times \frac{3}{4} = \frac{6}{20}$. This is smaller than *both* of the original numbers—multiplication makes nothing bigger here.

It is, however, worth thinking about how this relates to numerical calculations. When multiplying fractions, we multiply the two *denominators* (the numbers on the bottom) to get the resulting denominator—that's how many pieces a 1×1 square is divided into. And we multiply the numerators (the numbers on top) to get the resulting numerator—that's how many pieces end up shaded.

$$\frac{2}{5} \times \frac{3}{4} = \frac{2 \times 3}{5 \times 4} = \frac{6}{20}.$$

If you learned 'denominator' and 'numerator' as fancy-sounding words that didn't mean much, you might like to note that *denominator* is like *denomination*, which is appropriate because it's about how big the underlying fractional 'pieces' are. *Numerator* is like *enumerate*, which is about counting how many of those pieces we have. In this case, six pieces, each of 'denomination' $\frac{1}{20}$.

Now, the fractions so far have all been smaller than 1. But the area representation copes just fine with fractions bigger than 1. People tend not to think about those, because they think of fractions as 'small'. But $\frac{9}{4}$ ('nine quarters') is a perfectly reasonable fraction. It feels different, though, and mathematicians respect the distinction by calling a number of the form $\frac{m}{n}$ a *fraction*—or, as discussed in Chapter 5, a *rational number*—and such a number that is also less than 1 a *proper fraction*. The following calculation involves a proper fraction and an improper fraction. In this case, the resulting denominator is $3 \times 4 = 12$; each 1×1 square is divided into 12 equal pieces. The resulting numerator is $2 \times 9 = 18$; the number of shaded pieces is 18. So the result is $\frac{18}{12}$, which is equivalent to $\frac{3}{2}$. There is more on fraction equivalence in Chapter 3. In the meantime, can you imagine shunting around the shaded pieces so that they clearly occupy one-and-a-half of the 1×1 squares?

$$\frac{2}{3} \times \frac{9}{4} = \frac{2 \times 9}{3 \times 4} = \frac{18}{12}$$

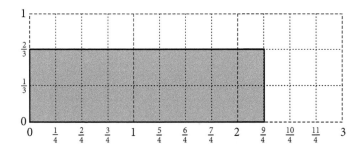

Next, consider $\frac{4}{5} \times \frac{5}{4}$. In this case, the resulting denominator is $5 \times 4 = 20$, and the resulting numerator is 4×5, which is also 20. The result, $\frac{20}{20}$, is equal to 1. In the diagram below, can you imagine shunting around the shaded pieces so that they clearly occupy a single 1×1 square?

$$\frac{4}{5} \times \frac{5}{4} = \frac{4 \times 5}{5 \times 4} = \frac{20}{20}$$

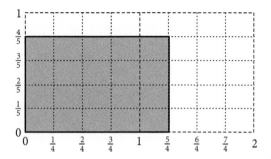

When two numbers multiplied together give 1, we say that they are *reciprocals* of one another ($\frac{5}{4}$ is the reciprocal of $\frac{4}{5}$ and vice versa). What other pairs of numbers are reciprocals? What would area diagrams look like for some of those pairs?

To conclude this section, I want to comment on the way that mathematics is sometimes explained in relation to 'real world' problems. Fractions are often represented using round things, which are supposed to be cakes or pizzas.

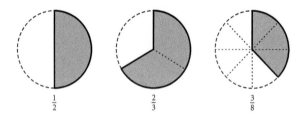

There's nothing wrong with this and, if you happen to have two adults and five children in the room, it is useful to be able to cut a cake into sevenths (I once cut a cake into respectably fair seven*teenths*—I was very proud of that). Fractions of round things are also handy for clocks. Knowing what a quarter-circle looks like means that I can 'see' a quarter of an hour in various positions on a clock face.

And circles can represent improper fractions; we just might need more than one cake.

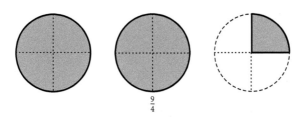

$$\frac{9}{4}$$

But circles have some limitations when compared with rectangular areas. Circles don't combine well with fractions as represented on number lines—there is no obvious 'zero' point and, even if we assign one (at 12 o'clock, say), an increase corresponds to sweeping out a greater area by traveling around a circle, not along a line. And circles don't work well for multiplying fractions—it is not clear how to use them for that. Different representations, as I've commented before, highlight different things. I think it's useful to be aware of their value not only for representing single items, but also for supporting valid reasoning about multiple items.

1.5 Squares

We'll return now to rectangles, and in particular to squares (a square is a *special case* of a rectangle—more on that in Chapter 2). We'll also work less with specific numbers and more with general notation.

First, think about areas of squares. A square with sides of length 3 has area 3×3. We can write 3×3 as 3^2 ('three squared'—not a coincidence), so a square with sides of length 3 has area 3^2. Similarly, a square with sides of length 8 has area 8^2, a square with sides of length a has area a^2, and a square with sides of length b has area b^2.

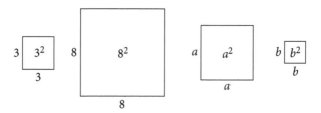

Now, what is the area of a square with sides of length $a+b$? The diagram below helps. The big square's sides have length $a + b$, so its area is

$$(a + b)^2 \ = \ (a + b)(a + b) \ = \ a^2 + ab + ba + b^2 \ = \ a^2 + 2ab + b^2.$$

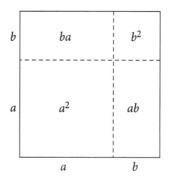

If you haven't seen this for a long time (or ever), make sure you can see how the algebra corresponds to the diagram—doing so is important for the rest of this chapter. Then note that I have stopped using the '×' symbol—I wrote 'ab' instead of '$a \times b$'. I think that's reasonable when using letters because, depending on your handwriting, an '×' can easily look like an 'x', which might be confusing. Also, I used commutativity without comment, turning a ba into an ab and adding it to the other one. People who are comfortable with algebra might not have noticed that—when commutativity is ingrained, it can be hard to see it in action.

Now, did you initially answer incorrectly, saying automatically that $(a+b)^2 = a^2 + b^2$? Many people do, and it's not their fault. The equation $(a+b)^2 = a^2 + b^2$ just looks like it ought to be valid, doesn't it? It's nice and tidy, and very similar to $2(a+b) = 2a + 2b$. But $(a+b)^2 = a^2 + b^2$ is not valid, because squaring is structurally different from multiplying by 2. A straightforward numerical check can highlight the problem. For instance, setting $a = 3$ and $b = 8$ gives

$$(a+b)^2 = (3+8)^2 = 11^2 = 121,$$

but

$$a^2 + b^2 = 3^2 + 8^2 = 9 + 64 = 73.$$

Obviously that's different.

I like the area representation better, though. It shows not only *that* $(a+b)^2$ does not equal $a^2 + b^2$, but also why it does not and how to fix it. In the following diagram, $3^2 + 8^2$ counts only the white areas, missing out the two grey rectangles.

$$(3+8)^2 = (3+8) \times (3+8) = (3 \times 3) + (3 \times 8) + (8 \times 3) + (8 \times 8).$$

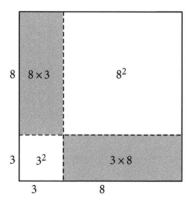

Similar representations can be used to think about more variables or more dimensions. Below is a diagram for $(a+b+c)^2$.

$$(a+b+c)^2 = a^2 + b^2 + c^2 + 2ab + 2ac + 2bc.$$

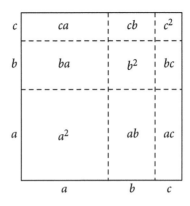

And here, adding a dimension, is a picture for $(a + b)^3$ ('a-plus-b cubed').

$$(a + b)^3 = a^3 + 3a^2b + 3ab^2 + b^3.$$

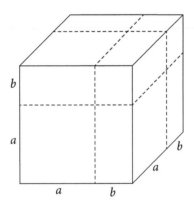

The 'pieces' are now three-dimensional, and some of them are 'round the back' in this two-dimensional drawing. Can you imagine holding the cube and pointing to each of the pieces? If you find that difficult, maybe make a physical version out of a box. If you find it easy, work out what the box would look like for $(a + b + c)^3$.

This section has introduced three equations:

$$(a + b)^2 = a^2 + 2ab + b^2;$$
$$(a + b + c)^2 = a^2 + b^2 + c^2 + 2ab + 2ac + 2bc;$$
$$(a + b)^3 = a^3 + 3a^2b + 3ab^2 + b^3.$$

All three are known in mathematics as *identities*, meaning that they hold for every possible combination of numbers. This distinguishes them from more general equations, which might hold for some numbers but not others. For instance, you might have been asked to solve equations like

$$2x + 9 = 3x + 6.$$

Such an exercise is meaningful only because this equation does *not* hold for every possible combination of numbers. If it did, there would be nothing to do—we could just say 'x is any number you like'. But this equation does not hold for $x = 10$, for instance (it would give $29 = 36$, which isn't right), or for $x = 1$ (it would give $11 = 9$). Solving it means finding the single value of x for which the equality is valid. In this case, that value is $x = 3$.

In Chapter 4 there will be more on equations like $2x + 9 = 3x + 6$. Here I want to observe that an identity, which *does* hold for all numbers, is a special kind of equation. To indicate this, mathematicians sometimes write identities using an extra bar on the equals sign. The first of the equations below could be read aloud as 'a-plus-b squared is identically equal to a-squared plus two-a-b plus b-squared.'

$$(a + b)^2 \equiv a^2 + 2ab + b^2;$$
$$(a + b + c)^2 \equiv a^2 + b^2 + c^2 + 2ab + 2ac + 2bc;$$
$$(a + b)^3 \equiv a^3 + 3a^2b + 3ab^2 + b^3.$$

I'll end this section by introducing one more identity that we'll use later. Here it is:

$$x^2 - y^2 \equiv (x - y)(x + y).$$

This is sometimes referred to as 'the difference of two squares', because of the $x^2 - y^2$ on the left. If you like algebra, you can check its validity by multiplying out, again using commutativity.

$$(x - y)(x + y) \equiv x^2 + xy - yx - y^2 \equiv x^2 - y^2.$$

If, like me, you like diagrams, it's possible to construct one for this, though it takes some dynamic imagination. Here's one way to do it. Start with a square with side length x, and imagine cutting out a smaller square with side length y. The remaining L-shaped bit has area $x^2 - y^2$:

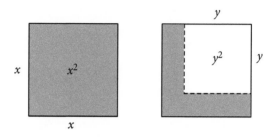

Now split the L-shape into two rectangles as shown in the following diagram. Imagine rotating and moving the smaller rectangle to line up with the bigger one; this works perfectly because they both have short side $x - y$. Moving the small rectangle does not change the total shaded area, and the long side of the resulting rectangle is $x + y$. Hence $x^2 - y^2$ is equal to $(x - y)(x + y)$.

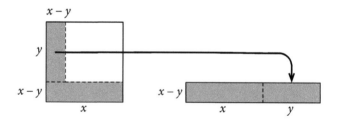

I'm abusing the diagrams now, though, so it is time to draw attention to their limitations. The diagrams for $x^2 - y^2 \equiv (x - y)(x + y)$ assume that y is smaller than x. But the identity still holds if y is bigger than x, and the diagrams don't represent such cases. More seriously, like all the diagrams in this chapter, they assume that all the numbers are positive. But the identities we've looked at—and commutativity and distributivity—hold for all numbers, not just positive ones. This means that the diagrams provide insight, but only up to a point. When a is negative, it's hard to represent $(a + b)^2 \equiv a^2 + 2ab + b^2$ with an area diagram. This makes some people uncomfortable, because they think that we shouldn't use a diagram that is not really 'good enough'. Certainly, they think, we shouldn't give the impression that these diagrams 'prove' that an identity holds for all numbers. Clearly that is true. But I find the diagrams useful anyway because they provide insight. That insight might be imperfect, but I think it's fine to use diagrams providing we keep an eye on their limitations. Throughout this book I'll point out similar issues.

1.6 Triangles

Most areas we've looked at so far have been rectangular. But we can get information about areas of other shapes by splitting up rectangles in new ways. Here is a diagram showing one way to do that, cutting a rectangle in half along a diagonal. This gives two triangles, each of which has half the area of the rectangle.

To focus instead on a given triangle and to find its area, people often label the base b and the height h, then imagine the 'other half' of the corresponding rectangle. This shows that the triangle's area is $\frac{1}{2}bh$ ('half times the base times the height').

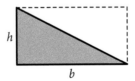

This diagram is simple because the triangle is right-angled so one corner fits neatly into the rectangle. What if that were not the case?

You might remember that the same formula still gives the area. Do you remember why?

It is possible to see why by taking the base of the triangle as one side of a rectangle, drawing the rest of the rectangle so that it 'fits' perfectly, then drawing a line from the top point of the triangle straight down to the base ('dropping a perpendicular'). This splits the diagram into two smaller rectangles, each of which contains its own triangle. Each mini-triangle takes up half of its corresponding mini-rectangle. So, adding the areas together, the whole triangle must take up half of the whole rectangle.

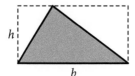 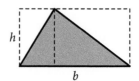

Dropping the perpendicular helps to focus attention on a useful split, and such focusing is important because sometimes it is hard to notice the right things. I have a vivid memory of a primary school teacher holding up some big cards with shapes on them. First, she held up a triangle like this.

Then she turned the card around so that the triangle looked like this.

Then she asked whether it was still a triangle. I said yes, but I was pretty much alone in this view—most of the other kids said it wasn't a triangle any more. My point is that when you are little, it is not obvious what is a key property of a mathematical object and what is incidental. If every triangle you see has a horizontal base—and we all tend to draw them that way because they look like they'll fall over if we don't—you might infer that a shape without a horizontal base is not a triangle. It's important to remember this when dealing with little kids, and it's important to recognize that no one grows out of the tendency to overgeneralize. For some mathematical concepts, you might retain slightly misguided ideas or fail to think about atypical cases. For instance, I just presented a diagram like this and invited you to agree that the area of a triangle will always be $\frac{1}{2}bh$.

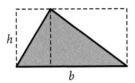

But is that true? Does the reasoning work for a triangle like the one below? Putting a rectangle around this gives a picture that is qualitatively different: the area of the triangle is less than half that of the rectangle.

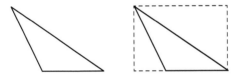

This seems to be a problem, though perhaps not a terrible one—we could get around it by treating a different side as the base.

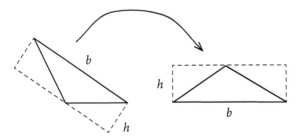

But is that necessary? In a given triangle, are there some sides that can act as the base and some that can't? Does choosing the 'wrong' base mess up the formula so that area can't be calculated in the same way? It feels to me like it shouldn't—there's no obvious reason to think that some sides of a triangle are 'special'. I'd like to be able to treat any side as the base, define the height as usual, and still find the area using the formula $\frac{1}{2}bh$. Can I do that?

The answer is yes, but it takes more thought to see why. First, it helps to distinguish the base of the triangle from the corresponding side of the rectangle, perhaps by labelling both the base b and the remaining bit of horizontal length—I'll call that a. Then we can follow the chain of reasoning captured in the diagrams that follow. The top two diagrams involve the areas of right-angled triangles. The third finds the area of the 'wonky' triangle by subtracting the second area from the first.

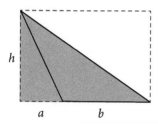

shaded area $= \frac{1}{2}(a+b)h = \frac{1}{2}ah + \frac{1}{2}bh$

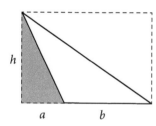

shaded area $= \frac{1}{2}ah$

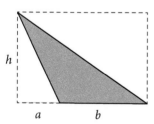

shaded area $= \frac{1}{2}ah + \frac{1}{2}bh - \frac{1}{2}ah = \frac{1}{2}bh$

All the diagrams require care in labelling the height—the direction of measurement must be perpendicular to the base, whether or not that coincides with another side of the triangle. But the general result is even better than it looks. It means that if we fix a base and a height, then slide the top point along at that height, the triangles generated all have the same area.

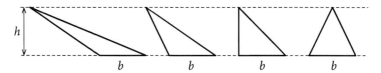

I don't think I'd have known that just by looking. The skinny triangle on the left is a lot longer than the squat one on the right, so I could have been convinced that its area is bigger. On the other hand, it is a lot skinnier, so I could have been convinced that its area is smaller. The fact that these things balance out perfectly, so that the areas are the same, is sort of thrilling for me.

And mathematics is full of similarly elegant results about invariance. But it's not just the results that have value. It would be a mistake to memorize the formula and forget the reasoning that led to it: the reasoning has power because it shows why the formula always works. Even better, it can be adapted to related situations. For instance, does the same kind of thing hold for quadrilaterals? Suppose we fix a base and height, then draw in a quadrilateral. What choices do we have? Obviously changing the length of the 'top' could change the area.

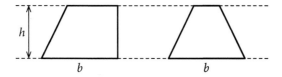

But what if we fix the top length—call it t, say—and just slide it along? Is the area of the quadrilateral invariant under this transformation? Can you convince yourself that it is, perhaps by dividing the shapes into simpler 'pieces' like rectangles or triangles? Are there multiple cases to consider, or can you construct an argument that covers all possibilities?

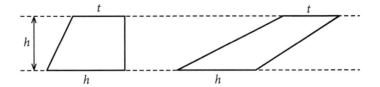

1.7 Pythagoras' theorem

Having discussed areas and triangles, we're finally ready for Pythagoras' theorem. Most people have hard of this, and many, when asked what it says, say something like

$$\text{'Um, } a^2 + b^2 = c^2\text{?'}$$

That's fair enough in a nonexpert. But even mathematics undergraduates do it, I think because school mathematics involves lots of exercises, which can encourage a focus on using formulas rather than on understanding what they mean. Pythagoras' theorem is about right-angled triangles, and specifically about a relationship between the lengths of their sides. The equation above should be related to a triangle labelled like this,[5] where the long side, of length c, is called the *hypotenuse*.

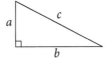

Then the theorem can be stated like this.

Pythagoras' theorem: A triangle is right-angled if and only if the square of its hypotenuse is equal to the sum of the squares of its remaining two sides.

To understand this, think about the relationship between the theorem, the formula, and the following diagram.

---

[5] What does the little square in the corner of the triangle mean?

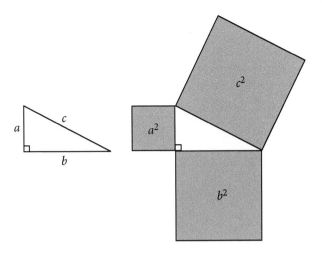

To check your understanding, ask yourself whether we could swap a and b in the formula. Could we? How about b and c? More importantly, do you believe Pythagoras' theorem? Does the area of the square on the hypotenuse seem to be about the same as the combined areas of the other two squares? I ask this not to make you doubt the theorem—I'm not about to tell you that there's a weird case for which it's not true. But people often learn mathematics without really thinking about how amazing it is, and I want to make sure that doesn't happen here.

Note that the areas do look about right—it's fairly easy to convince yourself that $a^2 + b^2 \approx c^2$ (the symbol '$\approx$' means 'is approximately equal to'). But the theorem doesn't say that $a^2 + b^2 \approx c^2$, it says that $a^2 + b^2 = c^2$; the equality is exact. And it says that this works for *every* right-angled triangle—no matter how skinny or squat. I don't think I'd have noticed that. I reckon I could have looked at thousands of right-angled triangles without once thinking about a general relationship between the squares of their sides.

So Pythagoras' theorem is pretty amazing. And, as with any theorem, our next question should be, 'Why is it true, then?' Mathematical relationships are not arbitrary: they exist for good reasons. Here is an explanation for Pythagoras' theorem, building on the ideas from previous sections.

We'll start with a 'big' square with side length $a + b$, and divide it up as before so that one smaller white square has area a^2 and one has area b^2. This time we'll also divide the two rectangles into triangles. Note that all four triangles are exactly like that shown to the right. Each is right-angled and each has one side of length a and one of length b; again we'll call the hypotenuse c. You can calculate the areas of the triangles if you like but, for the purpose of understanding why Pythagoras' theorem is true, their actual area doesn't matter.

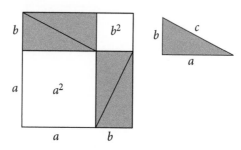

Now take another big square with side length $a + b$, and four triangles like those above. Arrange the triangles within the square in this different way (the lengths add up correctly: we still have one a and one b along each side). Now, what is the area of the wonky square in the middle? And can you see why Pythagoras' theorem must be true?

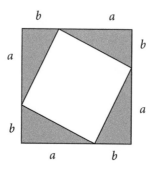

Below, for comparison, are both diagrams together. The argument is that the big squares are the same size so they have the same area. And the four triangles are the same size in both diagrams, so they have the same total area. So the areas that remain when the triangles are removed from the big squares must be the same. On the left, the remaining area is $a^2 + b^2$; on the right it is c^2. So we must have $a^2 + b^2 = c^2$.

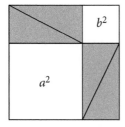 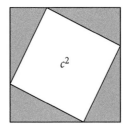

That's why Pythagoras' theorem is true. And that's the longest mathematical argument so far, so don't hesitate to re-read it if you want to. Also, as usual, there is more to think about. First, in the diagrams above, a and b are different. What if they were the same? What would the diagrams look like then? Is it easier to see that the theorem is true in that special case? Second, I didn't make a big deal of the fact that the triangles are right-angled. But right-angledness is essential, because Pythagoras' theorem is an *if-and-only-if* theorem. With the notation as above, both of these are true[6]:

$$a^2 + b^2 = c^2 \ \textit{if} \text{ the triangle is right-angled;}$$
$$a^2 + b^2 = c^2 \ \textit{only if} \text{ the triangle is right-angled.}$$

This means that Pythagoras' theorem provides a way of identifying right-angled triangles. We can tell whether a triangle is right-angled by labelling its longest side c and its other sides a and b, and checking whether $a^2 + b^2 = c^2$. If a triangle has side lengths 3, 4, and 5, for instance, then Pythagoras' theorem tells us that it is right-angled because

$$3^2 + 4^2 = 5^2 \quad (\text{check: } 9 + 16 = 25).$$

---

[6] Mathematically experienced readers might notice that the argument above only proves one of these. Which one? And everyone might like to know that Pythagoras' theorem can be proved in many different ways—try a simple internet search.

If a triangle has sides of lengths 4, 5, and 6, then Pythagoras' theorem tells us that it is not right-angled because

$$4^2 + 5^2 \neq 6^2 \quad \text{(check: } 16 + 25 \neq 36\text{)}.$$

(the symbol '$\neq$' means 'is not equal to'[7]). Because Pythagoras' theorem is valid, there is no need to draw triangles to check. Nevertheless, you might find that doing so helps to develop intuition. For the $(3, 4, 5)$ triangle, drawing is easy. Get hold of some squared or dotty paper[8] and draw one side of length 3 units and another at right-angle to it of length 4 units. The length of the hypotenuse will be 5 units.

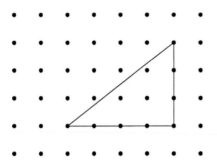

For the $(4, 5, 6)$ triangle it's harder because a non-right-angled triangle can't be lined up with pre-packaged squares. Some geometry helps, though. If you have a pair of compasses, here is one way to do it. First, pick your units (maybe centimetres) and draw one side. I'll do the side of length 6, making it horizontal. Then set the compasses to draw a circle of radius 4. Put the point at one end of the 6-line, and draw the circle. Note that every point on the circle is at distance 4 from its centre. So, wherever the third corner of the triangle is, it must lie on the circle.

---

[7] I really like the symbols '$\approx$' and '$\neq$'; I think they are just the ones that any sensible person would invent.

[8] When I was a kid this was a bit of a luxury. In the internet age you can download it.

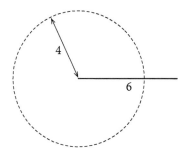

Next draw a circle of radius 5 centred at the other end of the 6-line. Because the third corner of the triangle must be on this circle too, it lies where the circles intersect.

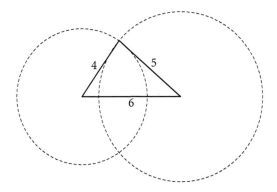

This shows that the angle between the sides of lengths 4 and 5 is not far off a right angle, which is unsurprising because $4^2 + 5^2$ is not far off 6^2 (6^2 is 'a bit too small', if you like). And this is the great thing about precise mathematics. We don't need to eyeball things and guess, we can check.

1.8 Pythagorean triples

The preceding arguments allow us to distinguish right-angled from non-right-angled triangles by squaring and adding their side lengths. In particular, $3^2 + 4^2 = 5^2$, so the *triple* $(3, 4, 5)$ corresponds to a right-angled triangle, and mathematicians call it a *Pythagorean triple*. Clearly, there are many different right-angled triangles and thus many different Pythagorean triples. But are there more *whole-number* Pythagorean triples, do you think? If there are more, how many? Just a few, or a lot, or perhaps infinitely many? And is there a way to find them all? Answers to these questions are fairly accessible, and you might want to put the book down and investigate for yourself. Can you find more Pythagorean triples, or convince yourself that none exist? If you have a computer, you might want to get a spreadsheet to do some of the work by calculating and adding combinations of square numbers. Do consider trying this before you read on.

One way to generate a new Pythagorean triple is to start with a known one and double the lengths. Because $6^2 + 8^2 = 10^2$ (check: $36 + 64 = 100$), $(6, 8, 10)$ is a Pythagorean triple.

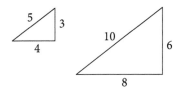

We could also triple the lengths, or quadruple them, and so on. So there are infinitely many Pythagorean triples. To mathematicians, though, this way of finding them is not very exciting. The resulting triangles all have the same proportions, so the bigger ones differ only in scale and not in a way that is mathematically interesting. Do there exist Pythagorean triples that correspond to differently proportioned right-angled triangles? To longer, skinnier ones?

People who've learned more mathematics sometimes know at least one:

$$5^2 + 12^2 = 13^2 \quad \text{(check: } 25 + 144 = 169\text{)}.$$

Is that it? If you investigated you might have found more, and if you used a spreadsheet you might have found lots. But you'll also have realized that this is not very satisfying. Experimenting on paper is hard work. Using a computer it is not hard work, but it does not provide much insight. If you used a spreadsheet and found 100 interestingly different Pythagorean triples, would you know for certain whether there are 100 more? Are there always more, and is there an efficient way to find them?

Here is one approach, using the *difference of two squares* identity from Section 1.5. We're interested in the equation

$$a^2 + b^2 = c^2.$$

I'd like to 'solve' this equation, but it is one equation in three unknowns so that is not directly possible. Also, I'm not keen on having all the variables tied up in squared terms. I can, however, get rid of some squares by subtracting b^2 from both sides and rewriting $a^2 + b^2 = c^2$ as

$$a^2 = c^2 - b^2,$$

then, using the difference of two squares identity, rewriting again:

$$a^2 = (c - b)(c + b).$$

To make further progress, it helps to specialize. Instead of trying to find *all* the Pythagorean triples, I'll try to find just some by temporarily making the problem simpler. One way to do this is to impose the restriction that $c - b = 1$; that is, that b and c are adjacent whole numbers. Then the equation reduces to

$$a^2 = c + b, \text{ where } b \text{ and } c \text{ are adjacent whole numbers.}$$

This looks more tractable. We can pick a square number and work out which (if any) adjacent numbers add up to it. For instance, 49 is a square number ($49 = 7^2$, so this corresponds to $a = 7$). Are there adjacent whole numbers that add up to 49? Yes. Because they are adjacent, one must be a little less than half of 49, and one must be a little greater, so $b = 24$ and $c = 25$ will work. This gives a new Pythagorean triple.

$$7^2 + 24^2 = 25^2 \quad (\text{check: } 49 + 576 = 625).$$

Running through the same reasoning for other square numbers leads to other Pythagorean triples. The following table shows results for the first ten square numbers:

a^2	a	b	c
1	1	0	1
4	2		
9	3	4	5
16	4		
25	5	12	13
36	6		
49	7	24	25
64	8		
81	9	40	41
100	10		

Note that only the odd rows of the table are filled in. Why is that? Note also that this method 'finds' the two known triples $(3, 4, 5)$ and $(5, 12, 13)$. It also finds two new, 'bigger' ones, $(7, 24, 25)$ as shown already, and $(9, 40, 41)$:

$$9^2 + 40^2 = 41^2 \quad \text{(check: } 81 + 1600 = 1681\text{)}.$$

By the way, I do not intend these checks as checks that the numbers come out right. I am totally confident that they will, because the reasoning is valid. I just quite like to see the actual numbers. If you feel the need for evidence because you're uncertain about the reasoning, try working through the paragraph about $a^2 = 49$ with some other square numbers.

Finally, note that the table also contains the Pythagorean triple $(1, 0, 1)$. This satisfies the requirements, because it is certainly true that

$$1^2 + 0^2 = 1^2.$$

But what has happened to the triangles? We can't have a triangle with a side of length 0, can we? Well, in one sense no, but in an other sense we could think of the triple $(1, 0, 1)$ as corresponding to a sort of extreme or *degenerate* triangle. Imagine taking a triangle with two sides of length 1, and making the angle between them smaller and smaller. In the limiting case in which the 'angle' is zero, the sides of the 'triangle' are 1, 0, and 1.

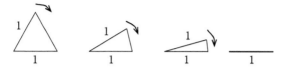

Some readers, quite reasonably, will think that's bonkers. Clearly, the degenerate triangle is not a triangle in the familiar sense. However, mathematicians are interested not only in 'typical' cases, but in general results about whole classes of cases. Sometimes a general result holds for typical cases and a degenerate one, so it can make sense to consider both together.

Whether or not you are up for that, I hope you are convinced that the above reasoning would generate infinitely many interestingly different Pythagorean triples. But does it generate them all? Probably not. Recall that I made the equation $a^2 = (c - b)(c + b)$ more tractable by introducing the restriction that $c - b = 1$. That restriction is quite hefty; there might well be Pythagorean triples for which $c - b \neq 1$. We can use similar reasoning again, though. What if $c - b = 2$, for instance? Then the equation

$$a^2 = (c - b)(c + b)$$

becomes

$a^2 = 2(c + b)$, where b and c are whole numbers that differ by 2.

So this time we want to take a square number a^2 and identify numbers b and c that differ by 2 and for which $a^2 = 2(c + b)$. For instance, 64 is a square number ($64 = 8^2$, so this corresponds to $a = 8$). Appropriate numbers b and c must add up to half of 64, which is 32. For them to differ by 2, one must be a bit less than half of 32 and one must be a bit greater—this time, $b = 15$ and $c = 17$ will work. This gives the new Pythagorean triple

$$8^2 + 15^2 = 17^2 \quad \text{(check: } 64 + 225 = 289\text{)}.$$

Again, similar reasoning can be applied with other square numbers:

a^2	a	b	c
1	1		
4	2	0	2
9	3		
16	4	3	5
25	5		
36	6	8	10
49	7		
64	8	15	17
81	9		
100	10	24	26

And again there are various things to note. Which rows do not get filled in this time, and why? Is there another degenerate case? Which triples are really new, and which are multiples of ones we already knew about? If the table were extended, could you predict which rows would contain new triples and which would contain multiples?

Finally, there is no need to stop there. We could let $c - b = 3$ in $a^2 = (c - b)(c + b)$. Or let it equal 4, or 5, or some other number. What happens in those cases? Which rows of the tables get filled in? Which contain familiar triples or multiples of familiar triples? If we carried on in this way, would we find all possible Pythagorean triples, or would this method miss some? The marvels of the internet mean that you can easily locate a lot more information about this. Probably more satisfying to give it some thought yourself, though.

1.9 Fermat's Last Theorem

Having explored Pythagoras' theorem and Pythagorean triples, we are finally ready for Fermat's Last Theorem. I hope you are now convinced that there are infinitely many Pythagorean triples, meaning—to draw together ideas from this chapter—that there are infinitely many interestingly different nondegenerate whole-number solutions to the equation

$$a^2 + b^2 = c^2.$$

For mathematicians, who like to generalize, this raises the question of whether there are also infinitely many interestingly different nondegenerate whole-number solutions to the equation

$$a^3 + b^3 = c^3.$$

The answer, I think, could easily be 'yes'. This equation involves cubing instead of squaring, so maybe finding appropriate triples (a, b, c) would take more work. But I'd have been quite prepared to believe that there are lots of them out there. In fact, however, there are not. There is not even a *single* nondegenerate whole-number solution. If you think about that properly, you'll see why mathematicians find it striking. The two equations look very similar but have dramatically different properties: $a^2 + b^2 = c^2$ has infinitely many solutions; $a^3 + b^3 = c^3$ has none.

Still more impressively, there are no nondegenerate whole-number solutions to

$$a^n + b^n = c^n$$

for any bigger whole number n, either. It's this result—that an equation of this form has no nondegenerate whole-number solutions for any n bigger than 2—that's known as *Fermat's Last Theorem*.

Fermat's Last Theorem is a favourite of mathematical expositors because it has a fascinating history. I will not describe that history here because the basic information is readily available and others have provided excellent detailed accounts. But I do want to round off this chapter by discussing the logical structure of Fermat's Last Theorem and the demands this makes on mathematical reasoning.

It can be quite straightforward to show that something *can* be done—we just need to give an example ('Look, here's a way to do it.'). But Fermat's Last Theorem is a claim that something *cannot* be done. An argument that something cannot be done must demonstrate that *no matter which numbers are chosen*, the equation of interest is not satisfied. Now, it is reasonable to wonder how that is possible. How could we ever know without checking? And how could we check infinitely many possibilities?

But the requirement can be formulated differently: the argument must demonstrate that whichever two whole numbers are chosen, their cubes (for instance) add up to a number that, for some reason, cannot be another cube. If you understand that, you understand the logical shape of Fermat's Last Theorem. You almost certainly wouldn't understand the accepted proof—the 'reason' in 'for some reason'—because it took 350 years of mathematical work and was only completed at the end of the 20th century. Although the theorem is fairly simple, the (current) proof is not. There's no reason to find that depressing, though, because most working mathematicians wouldn't understand it either. This surprises many people because, in school, they experienced mathematics as a set of procedures that appeared fixed and finished. They therefore tend to imagine that experts must know all the mathematics there is to know. They find it hard to imagine that there is so much that this is simply impossible, and even harder to imagine that new mathematics is still being created.

But mathematics, like other areas of human endeavour, is an evolving subject. And it is perhaps easier to think about this in historical context. At one time, people lived in primitive dwellings and had not invented systems for writing numbers, let alone procedures for multiplying fractions or for identifying right-angled triangles. These systems and procedures had to be developed, and when first developed they would have been cutting-edge mathematics accessible only to the privileged few—most people would have lacked the education to imagine their existence, never mind to understand their meanings. The equivalent is true today: fractions and Pythagoras' theorem are now taught in schools, but the frontiers of the subject continually move outward, and most of us find it hard to imagine what they must be like.

This book will not attempt to explain how today's creative mathematicians spend their time—we will get nowhere near the frontiers, as my aim is simply to explore and extend more basic relationships. But I will continue to highlight things that mathematicians attend to, things that they find interesting, and ways in which they think about concepts and formulate arguments. For that reason, I'll conclude each chapter with a brief review, drawing together its main points about conceptual understanding and mathematical thinking.

1.10 Review

This chapter took a speedy tour upward from basic multiplication, through areas, multiplying by fractions, algebraic identities, areas of triangles, Pythagoras' theorem, and Pythagorean triples, arriving finally at Fermat's Last Theorem. Throughout, I highlighted links between mathematical ideas and raised questions that readers might pursue.

This chapter also discussed broader mathematical ideas. It started with fundamental properties of the number system—commutativity and distributivity—and contrasted these with heuristics that are useful in some contexts but not universally valid (multiplication does not always make things bigger). I discussed identities like $(a + b)^2 \equiv a^2 + 2ab + b^2$, common related errors and ways to avoid these, and differences between identities and other equations. I discussed properties that are invariant under transformations like turning arrays around or sliding points along, or just 'looking' in different ways. I also discussed degenerate cases: atypically extreme versions of concepts. Finally, I talked about mathematically interesting differences. There are no absolutes here—different people find different things interesting—but some differences are more mathematically significant. I will continue to point these out.

In terms of mathematical thinking, this chapter discussed ways in which good representations can support insight. In particular, I'm a big fan of diagrams, and I used these to represent algebraic identities. But I also highlighted their limitations, which could encourage people to think only about simple or obvious cases. Some limitations can be overcome: adapting an existing argument showed that the area of a triangle can be calculated using the same formula regardless of its specific shape. But some can't: lengths are positive so they do not naturally represent negative numbers. Even serious limitations can sometimes be conquered using extra knowledge and imagination, but stretching the imagination will often sacrifice intuitive immediacy. It's useful to be aware of what representations do and do not do well.

Finally, I talked about mathematicians' habits of asking *why*, of finding representations that make mathematical relationships clear, and of building on useful reasoning. In particular, in a couple of places I said something like 'but we're not stopping there'. I think this is important because it's easy to be so taken with an interesting finding that you just stop, when there is a lot to be learned by extending the reasoning. These themes will recur in the remainder of the book.

CHAPTER 2

Shapes

2.1 Tessellations

Here is a *tessellation* or *tiling*.

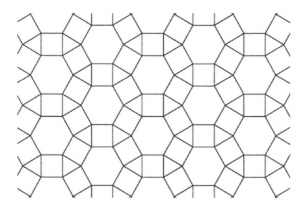

Many people find patterns like this beautiful due to their symmetry. How many different types of symmetry does this tessellation have? Pause now and count. You might want to consider *reflections*, *rotations*, and *translations*, and you might find this harder than you'd think—as in Chapter 1, there are decisions to make about which symmetries are really distinct.

Symmetry will be a central topic in this chapter, but we will begin by exploring four simpler properties of this tessellation and others like it. First, this tessellation is composed using *regular polygons*: equilateral triangles, squares, and regular hexagons. Such polygons are called *regular* because both their edge lengths and their *interior angles* are all the same (mathematicians say 'edges' where people more informally say 'sides'). Second, all the edges of all the polygons are the same length. Third, the polygons' edges are perfectly aligned. Fourth, at each vertex—*vertex* is the mathematical word for 'corner'—the same shapes are arranged in the same order. This is less obvious, but you can check by picking a vertex, finding the hexagon, and tracing around the vertex in a clockwise direction. The order in which the other polygons appear is always the same: hexagon, square, triangle, square.

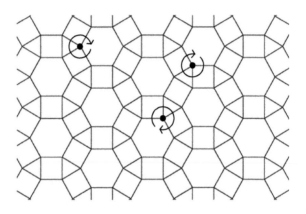

A tessellation with these properties is called *semi-regular*, and in this chapter we'll ask how many semi-regular tessellations exist. Think about this before you read on. Can you find more? Do you think there are lots more, or just a few? Might there be infinitely many, or do the conditions restrict the total number? If you've read Chapter 1, you'll notice that these questions are like those we asked about Pythagorean triples: how many things are there with certain properties, which ones are meaningfully different, and can we describe them all?

For semi-regular tessellations these are not trivial questions because there are lots of regular polygons: pentagons, heptagons, octagons, and so on. Maybe any and all of these can be used in semi-regular tessellations. But maybe not. Could we construct a semi-regular tessellation involving regular dodecagons, for instance (dodecagons are 12-sided)? Or one involving regular hectagons[1] (100-sided)? We will explore these questions, reviewing ideas about shapes and angles and developing an argument that we know how to generate all the possibilities. After that we'll look in more detail at symmetries. Before reading on, though, you might want to engage with the obvious challenge. How many semi-regular tessellations can you find?

2.2 Regular polygons

To think systematically about tessellations, it will help to review some ideas about regular polygons. First, we'll clarify the meaning of *regular*. I stated in the previous section that polygons are called regular when both their edge lengths and interior angles are all the same. But do we need both requirements? This depends upon the number of edges.

If a triangle's edges are all the same length, then its angles must all be the same. You might be able to 'feel' this by thinking about the *equilateral* triangle in the diagram below. Imagine grabbing one vertex and moving it so that an edge changes length. You'll find that you can't do that without changing some angles (how many angles must change?). Similarly, you can't change an angle without changing an edge length. Triangles are rigid: for triangles, saying 'the edges are all the same length' or 'the angles are all the same' amounts to the same thing.

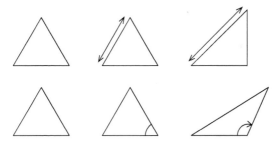

---

[1] I had to look up this word. For numbers of edges above about 12, mathematicians tend just to write the number and stick '-gon' on the end, as in 'an 18-gon', 'a 100-gon'.

For a four-sided polygon—a *quadrilateral*—this is not true. If a quadrilateral's edges are all the same length, its angles might all be the same, as in a square. But they might not. Squares are not rigid: we could squash a square by pushing on one of its corners. This would not change the edge lengths, but would transform it into a *parallelogram*, so called due to its pairs of parallel sides. In parallelograms like those shown below,[2] the angles are not all the same. Similarly, we could stretch a square to give a rectangle in which the angles are all the same but the edge lengths are not. So quadrilaterals can have four identical edge lengths without identical angles, or four identical angles without identical edge lengths.

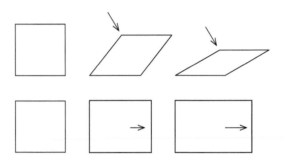

Something else to notice is that most people, looking at a square, would say, 'That's a square.' They wouldn't say, 'That's a rectangle' or 'That's a parallelogram.' But both statements would be true. A square is a rectangle because it is a quadrilateral with four right angles. It happens also to have identical edges, but that's an *extra* property—it doesn't stop it having four right angles so it doesn't stop it being a rectangle. Similarly, a square is a parallelogram because it is a quadrilateral with two pairs of parallel edges. It happens that those pairs are mutually perpendicular, but again that's an extra property. Think about this and you'll see that every rectangle is a parallelogram too, so squares sit within nested categories of quadrilaterals.

[2] As a child I learned to call a parallelogram with equal-length edges a *diamond*, but you don't really hear that in higher level mathematics.

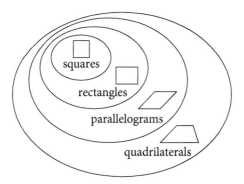

Children often have trouble with this nesting notion. They might feel quite strongly that a square isn't a rectangle. Adults often feel that too, but they can usually over-ride it because they are accustomed to everyday categorization. For instance, I'm typing this on a MacBook, which I could accurately describe as a MacBook, or as a laptop, or as a computer. Earlier I ate a Conference pear, which I could accurately describe as a Conference pear, or as a pear, or as a piece of fruit, or just as food. Similarly, a square is a rectangle, which is a parallelogram, which is a quadrilateral.

That said, calling a square a rectangle would be a bit rude, and calling it a parallelogram probably ruder. It wouldn't be a lie, but anyone doing that would be failing to give full information when they could do so in a simple way, so they'd be violating normal communicative conventions. These conventions are complex, though. Not long before eating the pear I said, 'I'm hungry, I need some food.' That didn't seem strange. But 'I'm hungry, I need a Conference pear' would have sounded pretty weird. In everyday life, categorization is a subtle business. In mathematics, systems of categorization are important, and concepts are often defined by starting with a pre-existing category and adding a restriction: 'a rectangle is a quadrilateral with four equal angles' is a perfect example.

The issue of defining becomes more salient when looking at hexagons, which are even more variable than quadrilaterals. A hexagon with six identical edges can be squashed in various ways, some of which have more symmetry than others.

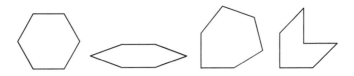

When you think about hexagons, probably you don't think about shapes like the three on the right. Probably you always think about regular hexagons, not these squashed kinds, which might therefore make you feel a bit twitchy. You might feel that although the middle two are hexagons, they are not very 'good' ones. They do at least stick out at all the corners, though—in mathematical terms, they are *convex*. The last is both squashed and 'dented'—you might think, is that a hexagon at all? The answer, according to mathematical convention, is yes. That shape is almost certainly not like your image of a prototypical hexagon, but it nevertheless satisfies the mathematical definition: it is *a closed plane figure with six straight edges*. Note, though, that if you still feel that the squashed hexagons are inferior, you're behaving quite mathematically. Mathematicians distinguish regular hexagons from other types because they have special properties.

Mathematicians also value clear communication, though, which is why definitions are needed. Mathematics involves deductive arguments, so we need to agree about which objects we're discussing—it is not practical to construct arguments about hexagons without deciding which things are hexagons and which things are not. And people don't always agree. Even if you wouldn't want to call the non-convex shape a hexagon, you can probably see that someone else might ('Well, look, it has six straight edges'). In this case, the collective decision is essentially to say, 'Yes, things like that will count as hexagons, but we will also define a subcategory called *convex hexagons* so that we can distinguish weirder hexagons from nicer ones; and we'll further distinguish our favourite ones by calling them *regular*.'

2.3 Regular tessellations

Some regular polygons can form *regular tessellations* involving polygons of only one type. Here is a regular tessellation using equilateral triangles.

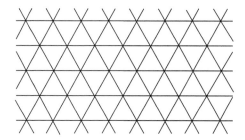

This has all the properties of a semi-regular tessellation. Here they are again.

- It is composed of regular polygons;
- All the polygons have edges of the same length;
- All the edges are aligned; and
- Around each vertex, the same polygons are arranged in the same order.

This means that a regular tessellation is a *special case* of a semi-regular tessellation: it has all the required properties so it is definitely semi-regular, and it has the extra property that it is composed using only one shape.

Squares, too, tesselate perfectly.

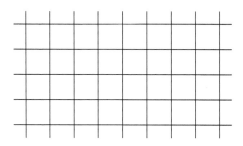

In both cases this occurs because the polygons have suitable interior angles. As you might recall, the angle 'around a point' is 360°. Equilateral triangles form a regular tessellation because their interior angles are 60°, and 60 is a *divisor* of 360. Squares form a regular tessellation because their interior angles are 90°, and 90 is a divisor of 360 too.[3]

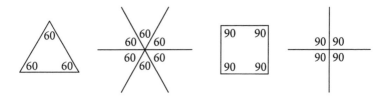

How about regular pentagons? Do they form a regular tessellation? No, because regular pentagons don't fit nicely around a vertex. In a regular pentagon, the interior angle is 108°, and 108 is not a divisor of 360. Three pentagons fit around a vertex (total angle 3 × 108° = 324°) but leave a wedge-shaped gap of 360° − 324° = 36°.

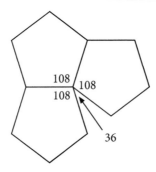

Regular hexagons do tessellate. Their interior angles are 120°, so three fit together perfectly.

---

[3] I have omitted the degree symbol '°' in the diagrams because it clutters them up.

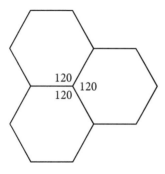

How about regular polygons with larger numbers of edges? Visual intuition might convince you that no others form regular tessellations. For instance, two regular octagons take up $2 \times 135° = 270°$, which leaves a gap of only $90°$—we can't fit another octagon in there. And for regular polygons with more edges, the gap is even smaller. So that's it—there are exactly three regular tessellations.

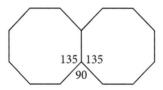

This thinking provides hints, though, about semi-regular tessellations using more than one type of polygon. Did you notice that between the two octagons, we couldn't fit another octagon but we could fit a square?

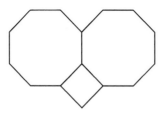

Starting with that and ensuring that the same shapes appear in the same order around every vertex leads to another semi-regular tessellation.

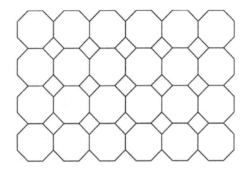

Hexagons and equilateral triangles also have potential, because removing a hexagon makes room for two triangles. This time, though, there are several possibilities for constructing a tessellation and it's easy to lose track of the list of conditions. Check for yourself—are these tessellations semi-regular?

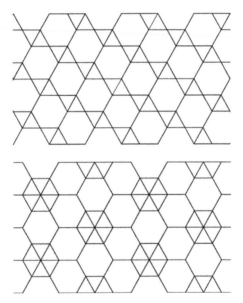

No, they're not. They're both tessellations and I think the bottom one especially is rather attractive—it has lots of symmetry, and I like the fact that I can see hexagons in it of three different sizes. But it is not semi-regular because it does not have the same shapes in the same order around every vertex. Around some vertices, the order is hexagon-hexagon-hexagon, around some it's hexagon-hexagon-triangle-triangle, and around the rest it's triangle-triangle-triangle-triangle-triangle-triangle. What are the equivalent lists for the top tessellation?

At this point we're up to five semi-regular tessellations: the three regular ones, the one with octagons and squares, and the one at the beginning of this chapter. Do you now want to change any of your answers to the questions in Section 2.1? How many semi-regular tessellations do you think there are? Do there exist semi-regular tessellations involving any regular polygon, or are some polygons no good for this purpose? Has the exploration so far changed your mind?

To generate more semi-regular tessellations, we could continue working out what shapes will fit around a vertex. But there are many possible combinations, so it's worth being systematic, and it would be useful to know all the relevant interior angles. So the next thing we'll do is calculate those.

2.4 Interior angles

One way to find interior angles is to imagine walking around a polygon. Imagine you are standing at the bottom-left vertex of the equilateral triangle in the following diagram and you walk along its bottom edge. When you get to the next vertex you are facing directly along the dotted line in the middle diagram. To walk along the next edge, you have to turn through the angle shown. Then you repeat this process until you get back to the beginning and turn once more to face in the original direction. Now, you have turned all the way around, so your total turn must have been 360°. You did it in three equal turns, so each turn must have been 360° ÷ 3 = 120°. The turn is not the interior angle, though, it is its *supplementary angle*. Because 180° is 'the angle along a straight line', each interior angle must be 180° − 120° = 60°.

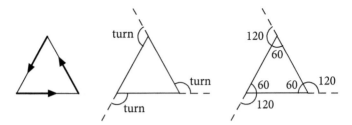

Having established that, there are two ways to find interior angles of other regular polygons. One is to generalize the argument, the other is to amend it then apply the amended result in a new argument. We'll do them in that order and compare the results.

To generalize, we'll start by applying the walk-and-turn idea to a regular pentagon. Walking around a regular pentagon still involves turning through 360°. But this time there are five turns, so each turn must be 360° ÷ 5 = 72°. So each interior angle must be 180° – 72° = 108°. Draw a sketch if you like, but don't worry about precision—it's quite hard to sketch a regular pentagon.

Then we can generalize fully. Suppose we have a regular n-gon (a regular n-sided polygon). I can't provide a picture of one of those and you can't draw one, because n is not a prespecified number. That doesn't matter for the written argument, though. For a regular n-gon, there are n turns. So each turn must be 360° ÷ n, meaning that each interior angle must be 180° – (360° ÷ n). I think that the expression 180° – (360° ÷ n) looks a bit ugly, so I'd prefer to reformat it and write

$$\text{interior angle of a regular } n\text{-gon} = 180° - \frac{360°}{n}.$$

This generalization gives a formula for calculating the interior angle for any given value of n. For instance,

$$\text{interior angle of a regular 10-gon} = 180° - \frac{360°}{10} = 180° - 36° = 144°.$$

Mathematicians like formulas, which save labour. Once we're convinced that the walking-and-turning argument always works, we don't need to keep rehearsing it—we can just stick a number n into the formula and do the calculation. If you like, take a moment to do that and confirm that the formula gives the expected answers for some familiar regular polygons.

But don't substitute evidence from calculations for understanding why the argument works—if you're unsure, apply the walking-and-turning directly to other regular polygons too. Here are some results in a table.

Number of Edges	Interior Angle	Number of Edges	Interior Angle
3	60°	8	135°
4	90°	9	140°
5	108°	10	144°
6	120°	11	147.27°
7	128.57°	12	150°

Not all regular n-gons have interior angles that can be measured in whole numbers of degrees; for those that do not, I've rounded to 2 decimal places. Note, though, that a lot of the simple ones do. That's because 360° is a good number to choose for the 'all the way around' angle: 360 has a lot of *factors*, meaning a lot of whole-number divisors. If you've ever wondered why we don't have 100 degrees in a circle, what do you think now? If it never occurred to you to wonder about such a thing, then note that this is a measurement question, and it did require a decision. Just as people eventually had to agree on standard units for measuring distance (miles, metres, and so on), they also had to agree on standard units for measuring angles.

Having generalized the walking-and-turning argument, we will now amend it and apply the amended result in a new argument. The amendment involves walking and turning for nonequilateral triangles. This is a bit harder because the turns are no longer identical. Some things don't change, though. The angle along a straight line is still 180°. So the total of the turns (call them t_1, t_2, t_3) plus the total of the interior angles (call them a_1, a_2, a_3) must add up to $3 \times 180° = 540°$.

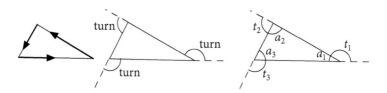

Also, because we still turn all the way around, the total of the turns $(t_1 + t_2 + t_3)$ is still 360°. So we must have

$$(t_1 + t_2 + t_3) + (a_1 + a_2 + a_3) = 540°$$
$$\Rightarrow \quad 360° + (a_1 + a_2 + a_3) = 540°$$
$$\Rightarrow \quad (a_1 + a_2 + a_3) = 540° - 360°$$
$$\Rightarrow \quad (a_1 + a_2 + a_3) = 180°.$$

Have you seen the '$\Rightarrow$' arrow before? It means 'implies' or (where it makes more sense) 'which implies that'. Try using that to read aloud the sentence that includes the equation array, from 'So we must have...' to the full stop. You'll find that you can do this in a grammatical way. This sometimes surprises people, because they think of mathematical symbols as separate from written English. But mathematicians write in full sentences—any well-written mathematical argument can be read aloud.

This argument shows that the interior angles of every triangle sum to 180°. If you vaguely remembered that from school, do you now understand why, and could you explain it to someone else? If you remembered it well, have you also seen it demonstrated in one or more other ways, and can you reconstruct them? And have you seen it used to work with other polygons? We can take any convex polygon and divide it into triangles by joining vertices with straight line segments. The number of triangles required depends upon the number of edges of the polygon. A convex pentagon—regular or not—requires three (make sure you believe that it doesn't matter which vertices we join). The total of the interior angles of all three triangles is $3 \times 180° = 540°$. So the total of the interior angles of every convex pentagon must be 540°. For me, this is a nonobvious invariant. Convex pentagons can vary quite a bit, and I would not have known just by looking that they all have the same total interior angle. But, having established this, I'm confident to specialize: if a pentagon is regular, each of its interior angles must be $540° \div 5 = 108°$.

Similarly, a convex octagon—regular or not—can be split into six triangles. The total of the interior angles of all six is $6 \times 180° = 1080°$. So the total of the interior angles of every octagon must be 1080°. If an octagon is regular, each of its interior angles must therefore be $1080° \div 8 = 135°$.

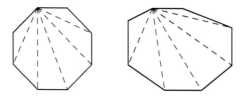

How many triangles do we need for an n-gon, and what does this tell us about the total of the n-gon's interior angles? If the n-gon is regular, what must each of its interior angles be? Can you convince yourself that the answer is $((n-2) \times 180°) \div n$? As a nicer looking formula, this could be written

$$\text{interior angle of a regular } n\text{-gon} = \frac{n-2}{n} \times 180°.$$

Now, one thing to note is that the two approaches—generalize-the-argument and amend-and-use-the-result—lead to different formulas for finding the interior angles of a regular n-gon. At least, the formulas look different—are they really? We'd better hope not. The interior angles of regular polygons are what they are, so the two formulas should give the same answer for every value of n. You could use calculations to convince yourself that they do for a lot of numbers, but there is no need: another great thing about general formulas is that we can use algebra to work with all cases at once. Here is a chain of equations showing that the second formula is, in fact, the same as the first:

$$\frac{n-2}{n} \times 180° = \left(\frac{n}{n} - \frac{2}{n}\right) \times 180°$$

$$= \left(\frac{n}{n} \times 180°\right) - \left(\frac{2}{n} \times 180°\right)$$

$$= (1 \times 180°) - \left(\frac{2 \times 180°}{n}\right)$$

$$= 180° - \frac{360°}{n}.$$

If you hesitated when reading the top line, try reading across the first equals sign from right to left instead of left to right. If you're still confused, note that this is a calculation involving $n$ths: n $n$ths minus 2 $n$ths is $(n-2)$ $n$ths, in the same way that n carrots minus 2 carrots is $(n-2)$ carrots. If you're still unsure, don't worry—we'll look at fraction addition in Chapter 3.

2.5 Mathematical theory building

I think it's worth a short interlude here to consider the two approaches taken in the previous section, because doing so illustrates ways to build up mathematical theory.

In the first approach, we began by generalizing the walking-and-turning argument from an equilateral triangle to a regular pentagon. That's not really generalization in the mathematical sense, though. It would better be called *application* because it involves applying a known argument to a new object (a new polygon, in this case). When mathematicians say 'generalization', they usually mean it in a more stringent sense. Ideally they'd like to demonstrate that an argument *always* works, so that there is no need to check individual applications. That's what we did next. By considering a regular n-gon, we generalized the argument to regular polygons with any number of edges.

To amend the argument we generalized in a different 'direction'. We fixed the number of edges at three but allowed the angles to vary, and thereby generalized to nonequilateral triangles. The result, that the interior angles of every triangle sum to 180°, was then available for use as an ingredient in a new argument. We used it to find the interior angles of polygons with more edges by splitting these into triangles, working out how many triangles are needed, and adding up their interior angles. Again, we first did a couple of applications, considering pentagons and octagons. Then we generalized to n-gons, this time without the requirement that these be regular. Having done that, we specialized back to regular polygons. We also checked that the formulas yielded by the two approaches are, in fact, the same.

I think informally about theory building in terms of 'levels', where going down a level is specializing, and going up a level is either generalizing or using an established result in a higher level argument. The theory developed so far used lots of simpler established ideas, too, such as information about the angle along a straight line. Such links between more basic and more advanced ideas lead successful reasoners to think of mathematics as a giant network of interconnected concepts.

2.6 Semi-regular tessellations

We can now list the interior angles of regular polygons. Here is an extended table to help in the quest to find semi-regular tessellations:

Number of Edges	Interior Angle	Number of Edges	Interior Angle
3	60°	14	154.29°
4	90°	15	156°
5	108°	16	157.5°
6	120°	17	158.82°
7	128.57°	18	160°
8	135°	19	161.05°
9	140°	20	162°
10	144°	21	162.86°
11	147.27°	22	163.64°
12	150°	23	164.35°
13	152.31°	24	165°

How might we approach that quest, though? Can you think of ways to work systematically? While drafting this chapter, I played around with a couple of strategies, and decided that the easiest[4] is to start with shapes with small numbers of edges, searching systematically for tessellations involving:

- Triangles only;
- Then squares and triangles;
- Then pentagons and squares and triangles;
- Then hexagons and pentagons and squares and triangles; and so on.

[4] There might be a better way, though, or you might prefer another one.

If there are infinitely many semi-regular tessellations, this search will go on forever but it won't miss anything. If there are finitely many possibilities, we will presumably realize that there is an inherent constraint—that it's possible for polygons to be 'too big' or similar. Which do you think will happen? What's your reason for that? And do you want to have a go before you read my reasoning?

Triangles-only we know about—that's a regular tessellation. So is squares-only. Squares and triangles we haven't yet explored. Starting with four squares at a vertex and removing one leaves a gap that can't be filled with triangles—one is too small and two won't fit. But removing two squares makes room for three triangles, yielding a new semi-regular tessellation (check).

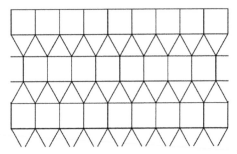

Does that cover it? In fact, no. Around each vertex above we have square-square-triangle-triangle-triangle. If instead we separate the two squares so that around each vertex we have square-triangle-square-triangle-triangle, that yields something different. I really like this new tessellation because I find it surprising—I wouldn't have thought that these shapes could make a tessellation so wibbly.

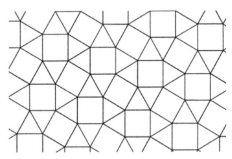

But there are no more tessellations with triangles and squares (check). So we're now up to seven semi-regular tessellations, and next on the search list is pentagons and squares and triangles. Do you feel optimistic about including pentagons? Probably not—they just don't feel like the right kind of shape, do they? (If you agree but you began this book without much confidence in your mathematical intuition, you might start to rethink that.) I wouldn't rule out the possibility that there's a tessellation using a pentagon plus polygons with a larger numbers of edges—maybe ten. But it doesn't seem likely that we'll be able to combine pentagons with squares and triangles.

We can demonstrate this by trying all possible combinations. Three pentagons around a vertex is no good because the gap is too small even for a triangle. How about two pentagons? Pentagons have interior angles of 108°, so the gap is now 360° – 216° = 144°. In that gap, one square will fit but without room for a triangle, two triangles will fit but with a gap, and all other possibilities are too small or too large.

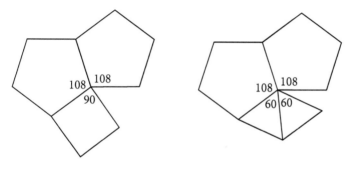

What if we have just one pentagon at a vertex? What angle remains then, and what combinations of squares and triangles might fit? Do any of them work? If you like drawing, try it. If you're less into diagrams, you might notice that we don't really need them—a numerical argument will do the job. Every vertex has total angle 360°, and 360° is a multiple of 10°. Triangles and squares have interior angles of 60° and 90°, which are also multiples of 10°. So, however many pentagons we use, those that meet at the vertex must have angles that add up to a multiple of 10. This isn't possible with one, two or three pentagons (108°, 216°, and 324°), and four won't fit. So no tessellations involve pentagons with squares, triangles, or both. I like diagrams a lot, but in this case we can search efficiently without them.

For hexagons we already have the hexagons-only tessellation and the hexagon-square-triangle-square one from the start of this chapter. But we can search as before. The previous observation again rules out pentagons because triangles, squares, and hexagons all have interior angles that are multiples of 10°. So we can start with three hexagons at a vertex, systematically remove one at a time, and fill the gap with squares and triangles. For some combinations there is more than one possible order, and remember that '$\neq$' means 'is not equal to'.

three hexagons

hexagon-hexagon-hexagon	$120° + 120° + 120° = 360°$

two hexagons

hexagon-hexagon-square	$120° + 120° + 90° \neq 360°$
hexagon-hexagon-triangle-triangle	$120° + 120° + 60° + 60° = 360°$
hexagon-triangle-hexagon-triangle	$120° + 60° + 120° + 60° = 360°$

one hexagon

hexagon-square-square-triangle	$120° + 90° + 90° + 60° = 360°$
hexagon-square-triangle-square	$120° + 90° + 60° + 90° = 360°$
hexagon-square-triangle-triangle	$120° + 90° + 60° + 60° \neq 360°$
hexagon-triangle-triangle-triangle-triangle	$120° + 60° + 60° + 60° + 60° = 360°$

Which rows of this table correspond to semi-regular tessellations that we've already seen? And which correspond to the two new ones below? I think the second of these has the quality of surprise again—it has a bit less symmetry than I would have expected.

Finally, which row of the table looks promising but does not yield a semi-regular tessellation? The table lists hexagon-square-square-triangle, for which the angles add up to 360°, so there is no problem there. Sketch

this configuration around a single vertex, though, and you'll find that there is a problem when we try to extend it: it's impossible to keep the same shapes in the same order around each vertex.

To conclude this section, what happens when we introduce regular heptagons (7-gons)? That's right—we can't use those. Their interior angles are not whole numbers of degrees, so they can't combine with polygons with lower numbers of edges—not even the pentagon. And they don't work on their own. For octagons we've already made a start: octagon-octagon-square is a possibility. Are there others? What if we start with one octagon at a vertex? What else, if anything, will fit? If there are promising configurations, do they yield semi-regular tessellations?

I'm going to pause here. If you're engaged by the investigation, you might want to try some more 'starting' shapes. If you've got the gist now and you want to know the full answer, read on, but be aware that the spoiler appears at the beginning of the next section.

2.7 More semi-regular tessellations

If you explored thoroughly, or perhaps strategically, you might have found two semi-regular tessellations involving dodecagons (12-gons). Both of these are shown below. Pretty, no? I still prefer the surprising squares-and-triangles one, but these are certainly pleasing.

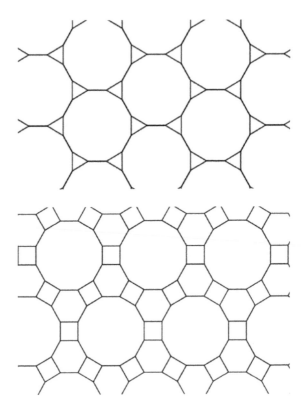

And that's it. You won't have found more semi-regular tessellations because there aren't any. There are just 11, including the three regular ones. How does that compare with your intuitive response at beginning of this chapter? Was your guess pretty close? Or were you a bit off, or perhaps miles off? More importantly, how do you feel now that you know the answer? If you feel satisfied, good. If, though, you feel a bit unsatisfied or deflated or let down, *that's even better*. If you've followed the arguments without too much difficulty, you should want to know not just *what* the answer is, but *why*. What renders it impossible to find more such tessellations?

If you read to the end of Chapter 1, you'll recognize this as a nonexistence question: it's about why there is no way to do something. In Chapter 1, I couldn't have explained why Fermat's Last Theorem is valid because the mathematics is too advanced. Here, a nonexistence argument still sounds difficult—what's to say that there isn't a semi-regular tessellation involving 1000-gons? But this time an argument is fairly accessible, and in this section and the next we'll demonstrate that polygons with more than 12 edges cannot be included in semi-regular tessellations. As usual, you might like to think before you read. What makes this impossible?

The key is to think about what goes wrong for regular polygons with large numbers of edges. As the number of edges increases, the interior angle gets closer to 180° (it can't go over 180° because all regular polygons are convex). Let's call the number of edges big N to emphasize the bigness, and say we start with an N-gon where N is 13 or more. The interior angles of this N-gon must be bigger than the interior angles of a 12-gon, meaning that they are bigger than 150°. That might not appear to constrain things much, but it does. Polygons sharing a vertex with the N-gon must have angles that sum to less than 360° − 150° = 210°. What possibilities are there? The smallest wedge we can put in the gap is an equilateral triangle. That knocks off another 60°, leaving a remaining gap of less than 150°.

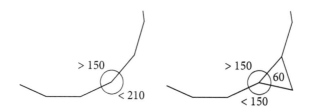

But this means that the 'biggest' regular polygon that can go with the N-gon is an 11-gon. So we only have to consider combinations from triangles 'up to' 11-gons. Here are the possibilities. Check that you believe I've not missed any.

11-gon-triangle	$147.27° + 60° = 207.27°$
decagon-triangle	$144° + 60° = 204°$
nonagon-triangle	$140° + 60° = 200°$
octagon-triangle	$135° + 60° = 195°$
heptagon-triangle	$128.57° + 60° = 188.57°$
pentagon-square	$108° + 90° = 190°$

This list is not overwhelming: just six possible combinations of 'other' shapes could go with our N-gon. If none of these works, then there are no a semi-regular tessellation with N-gons where N is 13 or more. So we just need to ask, for each of these possibilities, is there an N-gon with the right interior angles?

Although the 11-gon case appears first, I don't fancy tackling that because I'm worried about the nonwhole-number angle. So I'll start with the easier looking decagon-triangle combination. To fit with that, an N-gon would need interior angle $360° – 204° = 156°$. By reference to the table of interior angles, a 15-gon works. So we do have another possible combination. This doesn't yield a semi-regular tessellation, though. It might be hard to convince yourself by sketching—I can draw pretty accurate regular pentagons, but I'm totally defeated by a 15-gon. So I made a diagram as I did a lot of the others in this chapter, using a package called *GeoGebra*.[5] The 15-gon is really big, look.[6] Why exactly can't we extend this configuration?

[5] This is an open-source package so you can play with it for free—see http://www.geogebra.org.

[6] Before writing this chapter I had a vague idea that I'd make the polygon edge lengths the same in all the diagrams. That was never going to happen.

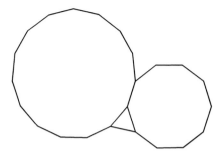

Similar reasoning about all four simple cases gives the following information about the angles needed for the N-gons and the corresponding values of N.

Other Shapes	Angles Total	N-gon Interior Angle	N
decagon-triangle	$144° + 60° = 204°$	$156°$	15
nonagon-triangle	$140° + 60° = 200°$	$160°$	18
octagon-triangle	$135° + 60° = 195°$	$165°$	24
pentagon-square	$108° + 90° = 198°$	$162°$	20

So there are more possible combinations, and they have a pleasing numerical elegance—nine-sided polygons with eighteen-sided ones and so forth (why does that happen?). The fact that the 20-gon yields a possible combination surprises but satisfies me because it shows that I was right to suggest that the pentagon might have potential. My guess that it might go with a decagon was wrong, but 20 is also a multiple of 5. Again, though, none of these combinations yields a semi-regular tessellation—think about why.

What about the messier possibilities?

$$11\text{-gon-triangle} \quad 147.27° + 60° = 207.27°$$
$$\text{heptagon-triangle} \quad 128.57° + 60° = 188.57°$$

As someone who's into pure mathematics, these make me nervous. I could do the same calculations but, because the tabulated angles are rounded to two decimal places, we've already lost precision. I'd be worried that rounding makes the calculations dodgy, so it might look like things don't match up when in fact they do. I'm not *that* worried—heptagons and 11-gons don't seem like they'll combine well with other polygons. But I've a mathematician's sense of tidiness (although I'm not actually a mathematician—more on that in the book's Conclusion) and I'd prefer to avoid thinking about rounding errors. Fortunately, this is another place where algebra is useful.

2.8 Algebra and rounding

We can work with awkwardly angled polygons without using the tabulated angle list. Instead, we can go back to the angle formula, keep all the numbers in fractional form, and put off calculating until the end. Here is the formula.

$$\text{interior angle of a regular } n\text{-gon} = \frac{n-2}{n} \times 180°.$$

And here's what I mean. For an 11-gon and a triangle, the total interior angle at the shared vertex is (in degrees)

$$\underbrace{\left(\frac{9}{11} \times 180 \right)}_{\text{11-gon}} + \underbrace{60}_{\text{triangle}}.$$

This means that the interior angle of an N-gon to go in the gap must be

$$360 - \left(\left(\frac{9}{11} \times 180 \right) + 60 \right).$$

Equating this with N-gon's interior angle gives

$$\frac{N-2}{N} \times 180 = 360 - \left(\left(\frac{9}{11} \times 180 \right) + 60 \right).$$

Now, that equation is a bit of a mess. Solving it won't be fun or satisfying, but there will be at most one solution because the equation combines one variable N with numbers. There are various ways to start, and pursued correctly any of them will lead to the same answer. But this is an opportunity to think before acting, checking first whether anything can be done to simplify the manipulations. For instance, the left-hand side has 180 as a factor, and 180 is also a factor of both 360 and $\frac{9}{11} \times 180$ on the right. It's not a factor of 60, though—that's a shame. But 60 is a factor of 180 and 360 and itself, and dividing through by 60 gives

$$\frac{N-2}{N} \times 3 = 6 - \left(\left(\frac{9}{11} \times 3 \right) + 1 \right).$$

That's no simpler structurally, but much simpler arithmetically. Next I'd tidy up the right-hand side so that it is just one fraction (if you're nervous about adding fractions, do read this bit but don't worry too much, and maybe come back to it after Chapter 3). The denominator 11 that appears on the right is not going anywhere because it has no factors in common with 9×3. So we'll make 11 the *common denominator* on that side. Observing that $6 = 66/11$ and $1 = 11/11$ gives

$$\begin{aligned}
\frac{N-2}{N} \times 3 &= 6 - \left(\left(\frac{9}{11} \times 3 \right) + 1 \right) \\
&= \frac{66}{11} - \left(\frac{27}{11} + \frac{11}{11} \right) \\
&= \frac{66}{11} - \frac{38}{11} \\
&= \frac{28}{11}.
\end{aligned}$$

Then solving the equation gives

$$\begin{aligned}
& \frac{N-2}{N} \times 3 = \frac{28}{11} & \\
\Rightarrow \quad & \frac{N-2}{N} = \frac{28}{33} & \text{(dividing both sides by 3)} \\
\Rightarrow \quad & N - 2 = \frac{28N}{33} & \text{(multiplying both sides by } N) \\
\Rightarrow \quad & 33N - 66 = 28N & \text{(multiplying both sides by 33)} \\
\Rightarrow \quad & 33N = 28N + 66 & \text{(adding 66 to both sides)} \\
\Rightarrow \quad & 5N = 66 & \text{(subtracting } 28N \text{ from both sides)} \\
\Rightarrow \quad & N = 66/5 & \text{(dividing both sides by 5).}
\end{aligned}$$

Whether or not you are confident about solving equations, I would encourage using words like those in the brackets. That's because it is tempting to think that letters can slide around in a magical way captured in slogans like 'change sides, change signs'. In my opinion, such slogans should absolutely not be used. I'm in favour of mnemonics for arbitrary connections. But I'm against them when connections are not arbitrary and when there is a simple, meaningful mathematical explanation. Here, what happens in all steps is that we *do the same to both sides of the equation*. We add the same thing to both sides, or subtract the same thing from both sides, or multiply both sides by the same thing, or divide both sides by the same thing. Because both sides are equal before, and we do the same to both, they remain equal. That's the main idea for solving equations of all types.

What does the result tell us, in this case? We started with the angles for a regular 11-gon and an equilateral triangle, and solved to find the number of edges N of a regular N-gon that will fit in their gap. But we don't get a whole number, meaning that this combination—11-gon and triangle and mystery N-gon—doesn't yield a possible tessellation. Try this for the final possible combination—heptagon, triangle, and mystery N-gon—and you'll find that doesn't work either. So we've now checked all the possible combinations of polygons that could go with a N-gon where N is greater than 12. None of these yield semi-regular tessellations, so we've established that there are no semi-regular tessellations involving polygons 'bigger' than dodecagons.

2.9 Symmetry: Translations and rotations

Another question at the opening of this chapter was about symmetry. Even those who don't find symmetry beautiful do tend to find it arresting, I think. We notice symmetry in the world, and we deliberately make things symmetrical for reasons of aesthetics as well as engineering. At the beginning of this chapter I asked about symmetries of the first semi-regular tessellation. How many did you find?

You might have found symmetries of all three types. The diagram below shows a dashed *line of reflection symmetry*, meaning (informally) that if you stick a mirror along the line and look into it from one side, you can again see the whole tessellation. If you haven't done this since you were little, I recommend trying it—interacting physically with mathematical concepts can be surprisingly engaging. The marked point with three arrows around it is a *centre of rotation symmetry*. To think about this, you might want some tracing paper. Trace over the whole tessellation then stick a pin in it at the centre of rotation symmetry and rotate your tracing around. Rotate it through 120° and everything will line up perfectly. Three such rotations return it to its original position, so this is a centre of *threefold rotation symmetry*. Finally, the straight arrow indicates a *translation* symmetry. Line up your tracing with the tessellation then shift the whole thing up and left according to the arrow; again it will match.

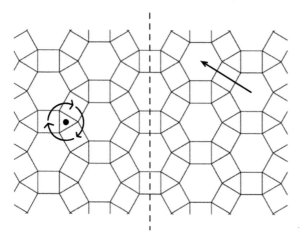

This section will consider questions about symmetries from a more advanced mathematical perspective, so we'll start not with this tessellation but with the simpler one involving octagons and squares. Also, although reflection symmetry is in some sense the most natural—it's what people often mean when they say that something is symmetrical—it's not the easiest to relate to an advanced perspective. Translations are simpler, so we'll start with those.

The octagons-and-squares tessellation has an obvious translation symmetry: slide everything one octagon to the right. And there are infinitely many similar symmetries: slide two octagons to the right, slide three octagons to the right, and so on. The original translation is said to *generate* all of these symmetries. Also, one could speak about the *inverse* of the translation as sliding everything one octagon to the left, and repeat this to generate others.

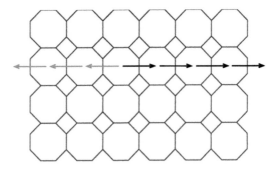

Then, of course, there are vertical translations. No amount of horizontal sliding will generate these. But again just one and its inverse will generate all the others. How about diagonal translations? Those too can be built from one-step horizontal and vertical translations. To get the diagonal translation below, for instance, we could perform two horizontal steps and one vertical one. In this way we can think of the horizontal and vertical one-octagon translations as 'building blocks': together they generate all translation symmetries for this tessellation.

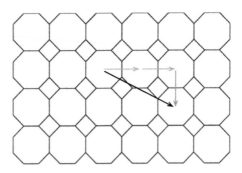

Such building-block or generator notions are common in mathematics. Mathematicians like to understand what is fundamental about mathematical situations, to strip them back to their most basic structures. Often that involves noticing that some objects or processes can be built by combining simpler ones, so that an entire structure can be understood in terms of a small number of components. Here we are building translations from simpler translations, but similarly we could build numbers from simpler numbers. For instance, numbers can be built from their *prime factors*, where 6 = 2 × 3, and 40 = 2 × 2 × 5, and so on (more on that in Chapter 5). Stripping back to building blocks requires abstract thought, so it's not always easy. Here, for instance, we have shifted our attention from the shapes that make up the tessellation to the translations that return it to itself. Translations are more abstract than shapes—we can't 'see' them in such a direct way—so they are harder to think about. If you want more practice, it might be worth playing around with the idea for other semi-regular tessellations. What are their translation symmetries? Do any of them have the same symmetries as this one, despite involving different shapes?

If you're ready to move on, we'll next ask how similar ideas apply to rotations. The centre of any octagon is a centre of fourfold rotation symmetry, as indicated below. Rotating through 90° about such a point returns the pattern to itself, as does rotating through 180° or through 270°; repeating the 90° rotation generates these others. Rotating four times through 90° takes the tessellation back to where it started. That's interestingly different from translating, notice. We could repeat the horizontal translation any number of times and never come back to the start. How do inverses work for rotations?

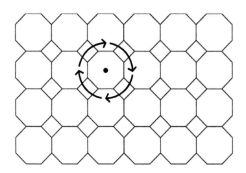

And how many rotation symmetries are there? Each octagon holds one, so there are infinitely many. But it turns out that we can build all of those by combining just this one rotation with translations. For instance, suppose that we want to rotate through 90° about the octagon marked with the cross in the following diagram. We can do that by translating according to the arrow in the top-left diagram to move the cross octagon to the centre of our existing rotation. That results in the configuration in the top right. Then we can rotate through 90° about the dot, resulting in the configuration in the bottom left. Then we can translate back according to the arrow in the bottom right. The configuration we end up with is exactly what we would get by performing a 90° rotation about the cross. In the diagram I've labelled some octagons to clarify where everything goes. But if the static diagrams don't convince you, get out some tracing paper, label or colour some more octagons, and try it for yourself.

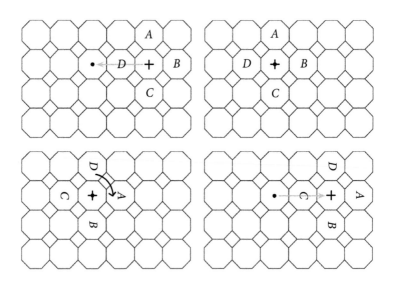

Similar processes will generate any octagon-centred rotation. But there are other rotation symmetries too. A 90° rotation centred in a square is different because it moves the surrounding octagons to their immediate neighbours, which the octagon-centred rotation does not. But again we could get any square-centred rotation by combining a single one with translations. Will just two rotations plus the translations generate all possible symmetries not involving reflections?

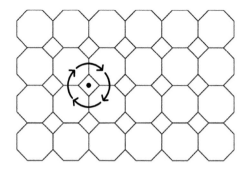

2.10 Symmetry: Reflections and groups

For the octagons-and-squares tessellation, one axis of reflection symmetry cuts the tessellation between two rows of octagons. But it's worth pausing here to consider the meaning of the word *symmetry*. When thinking about reflection symmetry, we tend to make observations, as in, 'The pattern is symmetrical.' But in the preceding section I wrote about symmetries as *transformations*—not as things we *observe*, but as things we *perform*. For translations and rotations this is pretty natural: we can imagine sliding or rotating the pattern (or the tracing paper over the pattern). For reflections it's less natural but still possible. We can imagine flipping the tracing paper over so that everything not on the axis swaps places with its mirror image.

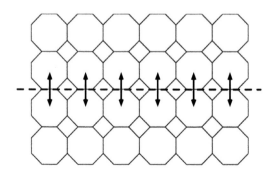

As a transformation, this reflection is its own inverse: flipping twice in the same axis returns everything to where it started. Is that true for every reflection? And can we build different reflections by combining this one with translations, or with rotations? Can you, for instance, work out how to generate a reflection in a different horizontal line, or in a vertical or diagonal line? This is another occasion for tracing paper—you might want to experiment.

If you do that carefully, you should find that we can combine this one reflection with translations and rotations to get numerous other reflections. Suppose, for instance, that we want to reflect in the vertical dotted line in the top-left diagram that follows. We can do that by performing our horizontal reflection in the dashed line (top right) then rotating through 180° about the point where the dotted and dashed lines meet (bottom left). That results in the configuration in the bottom right, which is exactly what we would get by performing the reflection in the dotted vertical line.

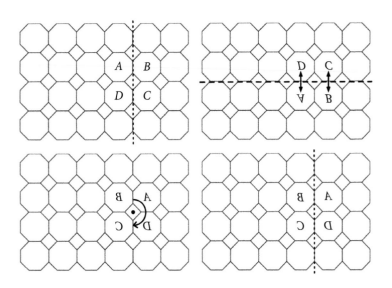

I could also explain how to get diagonal reflections, but it would develop more intuition if you experiment for yourself. Try it: draw an axis of reflection symmetry and work out how to build the corresponding reflection by combining the horizontal reflection with translations and rotations. Then, to convince yourself that this is possible in many cases, note that whatever combination of symmetries we perform, the pattern is either upside down or rightside up. If you're using tracing paper and the side you drew on ends up facing down, you need a reflection to get it that way. Much else can be done by translating and rotating.

If you are convinced of that and you want a challenge, you could investigate alternative generator combinations. For instance, what single symmetry would have the same effect as performing two reflections in distinct horizontal lines? What single symmetry would have the same effect as performing a horizontal reflection then a vertical one? If we had two nonparallel reflections and two nonparallel translations, could we generate all possible symmetries without using rotations?

At this point we are really working at an abstract level. We have shifted our attention from obvious objects (octagons and squares) through more abstract objects (symmetries of a whole tessellation) to the relationships between those abstract objects. Effectively, we're now doing arithmetic with symmetries, considering which ones we get by combining which others. If this is all new to you, you might like to know that we're thinking about *group theory*, an area of mathematics concerned with abstract structures. A mathematical *group* is a set of objects of some kind (in this case symmetries), together with a *binary operation*. Binary operations were discussed in Section 1.3, where I noted that addition is a binary operation: it takes two numbers (hence 'binary') and operates on them to give another (hence 'operation'). The operation for symmetries is *composition*, where *composing two symmetries* is a formal way of saying 'do one, then do the other'.

To form a group, the set of objects together with its binary operation must satisfy four *axioms*:

1. The group must be *closed*: using the binary operation to combine any two objects in the set must give another object in the set. Adding two numbers always gives another number. Composing two symmetries always gives another symmetry, because each replaces the tessellation perfectly on itself.

2. There must be an *identity*, an object in the set that does not change the others. When adding numbers,[7] the identity is 0 because $n + 0$ always equals n. When composing symmetries, the identity is *do nothing* (this might seem weird, but 'do nothing' is a perfectly good symmetry, just as 0 is a perfectly good number).

3. Every element must have an *inverse*, an object that 'undoes' its effect. When adding numbers,[8] the inverse of 2 is –2. When composing symmetries, the inverse of *rotate through* 90° is *rotate through* 270°, and so on.

4. The operation must be *associative*: if a, b, and c are objects in the set and the binary operation is $*$, we must have $(a*b)*c = a*(b*c)$. When adding numbers it is always true that $(x + y) + z = x + (y + z)$. When composing symmetries A, B, and C, it is always true that $(A \circ B) \circ C$ has the same effect as $A \circ (B \circ C)$ (the symbol '$\circ$' is often used for composition). This axiom is often the hardest to think about.[9]

Group theory is part of an area of mathematics called *abstract algebra*, which is an appropriate name because it is indeed very abstract. These axioms are not about numbers, or about individual addition sums, or even really about relationships between addition sums. They're about properties of the whole set of numbers and how they behave under addition. Similarly, the axioms are not about tessellations, or about individual symmetries of tessellations, or even really about the relationships between those symmetries. They're about properties of the whole set of symmetries and how they behave under composition. Nevertheless, in both cases, *the axioms are the same.* That's why they are worth stating, and that's why mathematicians study subjects like abstract algebra—they are drawn to cases in which superficially different things have the same underlying structures. Numbers and symmetries feel very different, but behave in remarkably similar ways.

---

[7] When multiplying numbers, the identity is 1 because $n \times 1$ always equals n. The identity depends upon the binary operation.

[8] What is the multiplicative inverse of 2?

[9] Not all binary operations are associative, just as not all binary operations are commutative. For numbers, it is not always true that $(x - y) - z = x - (y - z)$.

2.11 Symmetry in other contexts

In this chapter we have considered symmetries of infinite tessellations. But ideas about symmetries and groups can also be applied to other objects. Simpler cases involve symmetries of single shapes. What are the symmetries of a regular pentagon? Translations are irrelevant—we can't shift a pentagon to the right while landing it back on itself. But rotations and reflections are possible. How many symmetries does a pentagon have, and how many are needed to generate its symmetry group?

Symmetries can also be used to classify patterns. For instance, there are exactly 17 distinct *wallpaper groups*, meaning that every possible repeating-pattern wallpaper has one of just 17 underlying symmetry groups. The designs might look different—maybe your wallpaper has swirls and your child's has spaceships. But there are only 17 fundamentally distinct ways of arranging the swirls or spaceships on wallpaper.

We can also generalize to objects with more dimensions. Can you list all the symmetries of a cube? Again translations are irrelevant, but the extra dimension makes reflections and rotations more complex. For instance, we could reflect a cube not in a line but in a *plane* as labelled in the following left diagram. Are there reflection symmetries for which the plane cuts the cube in different ways? Through vertices, for instance, rather than faces and edges? Similarly, we could rotate a cube not about a point but about an *axis* dropped directly through its centre, as in the diagram on the right. Through what angles could we rotate it to give symmetries? And what is the corresponding answer if we put the axis from one vertex to its diagonal opposite?

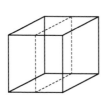 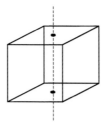

Finally, we can ask about symmetries for patterns like the one on the front of this book. That picture shows a *Penrose tiling*, named after the mathematician Roger Penrose. Here is a version in black and white so that you can inspect it carefully.

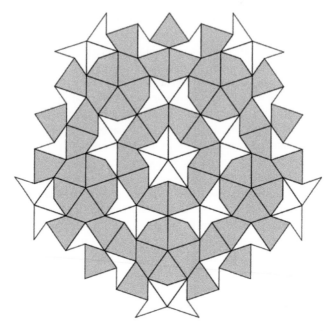

Note that the whole pattern is composed using just two types of tile, which are usually called *kites* and *darts*. The kites are convex and have interior angles 72° (three of these) and 144°. Darts, like the dented hexagon in Section 2.2, are not convex; their interior angles are 72°, 36° (two of these), and 216°.

If the 72° angle looks familiar, that's because it appeared in this chapter in relation to regular pentagons. And that's not a coincidence—Penrose tilings are intimately related to pentagonal symmetry. This particular tiling has fivefold rotation symmetry about its central point, and five axes of reflection symmetry. Does it have translation symmetry? That's less obvious. There's no visible repeating pattern *in the diagram*, but the diagram shows only a finite portion of an infinite tiling. Perhaps if we extended it, we would be able to identify a translation symmetry.

But you might have guessed that this is interesting because *no such translation exists*. Penrose tilings are *aperiodic*, meaning that they have no translation symmetry. The shapes fit together perfectly and the pattern extends forever, but it never repeats. Never. I think this is brilliant, and we are privileged to know about it: aperiodic tilings were first constructed in the 1960s, and Penrose tilings in the 1970s. There are kite-and-dart tilings with different central configurations—you might like to experiment with other ways to put the shapes together.[10] And there are aperiodic tilings composed using different shapes. All have extremely interesting properties, and you can learn more with a simple internet search. Finally, if you agree that this is brilliant and you would like to stand on some Penrose tiles, you can do so outdoors at the Mathematical Institute at the University of Oxford, and indoors at the Department of Mathematics at Texas A&M University. I hope there are more places too.

2.12 Review

This chapter began with questions about a single semi-regular tessellation. It then discussed polygons, regular tessellations, interior angles, searching systematically for semi-regular tessellations, symmetries, groups, and aperiodic Penrose tilings. As in Chapter 1, I highlighted links between mathematical ideas and raised questions that readers might pursue. For those who are interested, there is plenty to follow up.

I also once again discussed broader mathematical ideas. These included precise definitions and nested classifications. I highlighted both

---

[10] Searching for 'GeoGebra Penrose' yields a downloadable worksheet by Anthony Or that allows you to do this.

the labour-saving value of general formulas and the utility of algebra for constructing general arguments—for instance, for establishing the equivalence of two differently derived formulas. I also noted that mathematicians write in sentences. Mathematical arguments might contain both symbols and words, but they can be read aloud as one would read other text. As you continue with this book, I would encourage you to check that you are really *reading* the mathematics in this meaningful way.

This chapter also discussed mathematical theory. Theory was central in the later sections, which discussed abstract ways to examine tessellations, and the fact that different structures might have the same underlying group theoretic properties. But theory building also operates at more concrete levels. In Section 2.5 I commented that mathematicians are often less interested in *applying* mathematical arguments than in *generalizing* to establish that they always work. I think this contrasts interestingly with what many people experience during their mathematical education. For many, doing mathematics means doing applications—repeating standard calculations for different numbers or angles or equations. Applications can be satisfying, of course—a page of correct answers might bring pleasure, even if many people later decide that such work is boring. But, for me, the satisfaction of a job well done is quite different from the satisfaction that comes from understanding why an argument works. I find the latter much deeper and more gratifying. If you began this book believing that mathematics is repetitive and dull, are you starting to see what I mean?

CHAPTER 3

Adding up

3.1 Infinite sums

Have you heard the argument that it is impossible to leave a room? To get to the door you have to go halfway. Then you have to go half of the remaining distance, then half of that, and so on. Some distance always remains, so you can never get out of the room.

Presentations of this apparent paradox sometimes make me irritated. Not because the underlying issues are uninteresting—on the contrary, they encompass some beautiful mathematics. What irks me is that they sometimes encourage people to treat this argument as an isolated curiosity and, because everyone knows that they *can* get out of the room, to conclude that mathematics is nonsense.

Mathematics isn't nonsense, and considered carefully this is not a paradox: it is a question about *infinite sums*. To see this, it helps to introduce numbers. Suppose that the door is two metres away. To get to it, you must travel half of the distance (one metre), then half of what is left (half a metre), then half of what is left (a quarter of a metre), and so on. In total you must travel

$$1 + \frac{1}{2} + \frac{1}{4} + \frac{1}{8} + \frac{1}{16} + \frac{1}{32} + \dots$$

metres, where the ellipsis '...' means 'and so on forever'.

For some people the apparent paradox arises because the argument implicitly invites us to think that each successive fraction of the distance takes the same amount of time. If it did take, say, one second to go half way, and another to go half of what's left, and another to go half of what's left, then it would indeed take forever to get out of the room. It doesn't, though, so that's not really a problem. And in fact there is no problem at all, because $1 + \frac{1}{2} + \frac{1}{4} + \frac{1}{8} + \frac{1}{16} + \frac{1}{32} + \ldots$ is equal to 2; the distances added together cover the two metres. This can be represented on a number line.

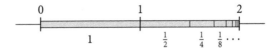

You might not be happy with that, though. You might think, 'Well, the fractions all fit, but the sum isn't 2 because there's always a bit left over.' If so, hold that thought—its mathematical resolution will be discussed in this chapter. For now, note that $1 + \frac{1}{2} + \frac{1}{4} + \frac{1}{8} + \frac{1}{16} + \frac{1}{32} + \ldots$ is an *infinite sum*— also known as a *series*—because it has infinitely many *terms* or *addends* (things that are added up). But its total is nevertheless finite. There are infinitely many terms, but their total is certainly not bigger than 2.

To complete this introduction, I want to raise a question about a different series:

$$1 + \frac{1}{2} + \frac{1}{3} + \frac{1}{4} + \frac{1}{5} + \frac{1}{6} + \ldots.$$

This is known as the *harmonic series*. It too has infinitely many terms and, as for the previous series, each term is smaller than its predecessor. The following diagram represents the sum of the first six terms (check that you believe I've made the bars the right lengths).

What do you think is the total of this infinite sum? It is certainly more than 2 because adding the first four terms gives $1 + \frac{1}{2} + \frac{1}{3} + \frac{1}{4} = \frac{25}{12}$ (don't worry if your fraction addition is rusty—we'll review that). Is the total bigger than 3? Bigger than 4? How big do you reckon it is overall? You might want to write down an estimate so that later you can see how close you were. As in the preceding chapters, we will work up to the mathematics needed for this by starting with more basic material, this time on fractions and addition.

3.2 Fractions

How do you think about fractions? There are various ways to do it, and they highlight different aspects of this type of number. To see what I mean, try this.

Which is greater, $\frac{2}{7}$ or $\frac{4}{7}$?
And again, which is greater, $\frac{2}{7}$ or $\frac{2}{9}$?
And again, which is greater, $\frac{2}{7}$ or $\frac{4}{11}$?

Did you find the second question harder than the first? And the third harder still? Most people do. In the first question, the second fraction has a bigger numerator and is therefore bigger. In the second question, the second fraction has a bigger denominator, and is therefore *smaller*. In the third question, both numerator and denominator are bigger, so the components don't lend themselves to straightforward comparison and the question demands more sophisticated, strategic thought. Mathematically educated adults tend to know all of this and to get the answers right, but in the later cases they are slower and more prone to error. It takes extra thinking to over-ride the instinct that fractions involving 'bigger numbers' will be bigger, and to invoke a reasoning strategy that takes account of overall magnitude rather than component sizes. And this is a robust effect: even professional mathematicians take longer to judge the relative sizes of more difficult fraction pairs, and they make more mistakes when judging under time constraints.

So, if you did find the second and third questions harder, you're in good company. This means that it is useful to represent fractions in ways that clarify their relative sizes. One approach is to use diagrams like those in Chapter 1, perhaps with circles.

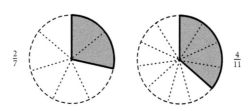

$\frac{2}{7}$ $\qquad\qquad\qquad\qquad$ $\frac{4}{11}$

This supports comparison, but only if the diagrams are accurate—it's not much use if you can't draw elevenths.

Another approach is to convert to decimals. For instance,

$$\tfrac{2}{7} = 0.285714\ldots \quad \text{and} \quad \tfrac{4}{11} = 0.363636\ldots.$$

This supports comparison because decimals use a standard scale of tenths, hundredths, and so on. For instance, $\tfrac{2}{7}$ is less than 0.3 and $\tfrac{4}{11}$ is greater than 0.3, so $\tfrac{4}{11}$ is bigger. Decimals raise other questions, though. Each of these *decimal expansions* is infinite, so in decimal form we can't write down the 'whole' number; writing $\tfrac{2}{7}$ is more economical and more precise. You might be inclined to observe that $\tfrac{4}{11}$ has a repeating pattern, meaning that it could be written economically as $0.\dot{3}\dot{6}$ or $0.\overline{36}$ (where the digits between the dots or under the line repeat forever). If so, though, you should ask yourself a question. How sure are you that the pattern repeats forever? How do you know that it doesn't change at the tenth digit, or the hundredth? Have a think about that—there'll be more on it in Chapter 5. For now we'll stick to fractions.

I often think about fractions as ratios of lengths. For $\tfrac{2}{7}$, I see in my mind's eye a bar of length 2 and another of length 7. The 2 is two sevenths of the 7, so I can 'see' $\tfrac{2}{7}$ as that ratio. To think of $\tfrac{2}{7}$ more explicitly as a number between 0 and 1, I can mentally rescale, which really just involves relabelling.

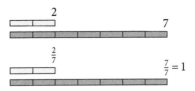

Rescaling also helps for thinking about equivalent fractions. For instance,

$$\frac{1}{3} = \frac{2}{6} = \frac{3}{9} = \frac{4}{12},$$

which all look the same in the following diagrams because each has the same numerator-to-denominator ratio. That's what it means for fractions to be equivalent.

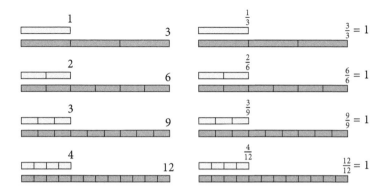

You might like to know that some languages write fractions explicitly as ratios. I learned this only recently because people tend to translate for international audiences. But Belgians, for instance, write 1 : 3 instead of 1/3. The English notation 1/3 highlights a division meaning for fractions, which I'll discuss in Chapter 5. The Belgian notation highlights the ratio meaning. Remember that in Chapter 1 I said mathematicians refer to fractions as *rational* numbers? Do you see why?

Ratio thinking supports my strategy for fraction comparison. I wouldn't want to convert to decimals because I can't easily do that in my head. And comparing $\frac{2}{7}$ with $\frac{4}{11}$ is difficult because these fractions are similar sizes and because my mental representation of sevenths and elevenths isn't great. The representations in this book are accurate, but that is because (for instance) I made the circles carefully using a computer package and properly calculated angles. My imagination is more like freehand drawing, and I wouldn't need to be off by much to be wrong. But my mental approximations are good enough for me to see that both $\frac{2}{7}$ and $\frac{4}{11}$ must be around $\frac{1}{3}$. Having recognized that, I can do easier comparisons, working out that

$\frac{2}{7}$ is a bit less than $\frac{1}{3}$, because $\frac{2}{7} < \frac{2}{6} = \frac{1}{3}$,

and that

$\frac{4}{11}$ is a bit bigger than $\frac{1}{3}$, because $\frac{4}{11} > \frac{4}{12} = \frac{1}{3}$.

So $\frac{4}{11}$ is bigger than $\frac{2}{7}$. I like chains of equations and inequalities so I might summarize this by writing

$$\frac{2}{7} < \frac{2}{6} = \frac{1}{3} = \frac{4}{12} < \frac{4}{11},$$

which would be read aloud as 'two sevenths is less than two sixths, which is equal to one third, which is equal to four twelfths, which is less than four elevenths'.

From representing ratios using bars, it is only a short step to number lines, which are hard to beat for size comparisons. Here is a number line equivalent of the $\frac{2}{7}$ diagram.

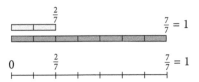

And here are some number lines showing fractions between 0 and 1 with different denominators.

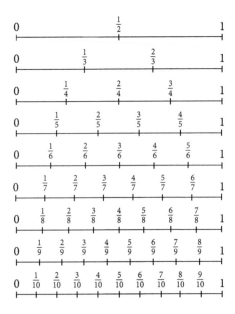

Number lines also highlight other ways in which rational numbers differ from whole numbers, where whole numbers are properly called *integers*. Consider this question.

How many numbers are there between $\frac{3}{8}$ and $\frac{7}{8}$?

The answer, of course, is infinitely many—you can identify some using the number lines above. But can you guess what incorrect answer children sometimes give? Yep, three. They tend to think

'three eighths, *four eighths, five eighths, six eighths,* seven eighths.'

Effectively they treat rational numbers as though they behave like integers. For some purposes that's okay. Like integers, rational numbers can be ordered on a number line. But, unlike integers, rational numbers can't be 'counted' in order from left to right. For any given rational number there is no 'next' number, because for any candidate next number, there's a number closer. If that makes you hesitate, try starting at $\frac{1}{2}$. The number $\frac{6}{10}$ is pretty close to this, but it's not as close as $\frac{11}{20}$, or $\frac{21}{40}$, or $\frac{501}{1000}$. However close we get, there are always more numbers in the gap.

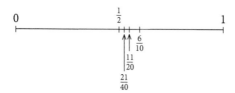

I alluded to this difference in a footnote in Chapter 1. The *natural numbers*—the counting numbers—can be thought of in terms of discrete objects, or in terms of marks or distances on a number line.

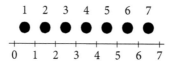

The rational numbers can't be thought of as discrete line-up-able quantities, although they can be represented on a number line, at least with some imagination. We can't draw infinitely many marks, but we can imagine zooming in to see more and more.

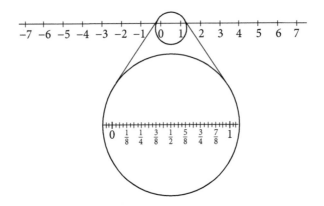

Taking that to its logical conclusion, we can think about the number line as a *continuum*; shifting from integers to rational numbers shifts from discrete towards continuous quantities. The rational numbers aren't, in fact, enough to fill the continuum, which I'll explain in Chapter 5. But the number line is useful for adding fractions, which we'll do next.

3.3 Adding fractions

If the thought of adding fractions strikes terror into your heart, you're not alone. For many people, this marked the point at which mathematics started to seem like meaningless symbol-pushing. One problem is that everyone has a sense of symbolic tidiness, so we're all a bit annoyed that

$$\frac{2}{3} + \frac{1}{8} \neq \frac{2+1}{3+8}.$$

After all, as discussed in Chapter 1, it *is* true that

$$\frac{2}{3} \times \frac{1}{8} = \frac{2 \times 1}{3 \times 8}.$$

Unfortunately, the addition version doesn't work. It can't, because the sizes are all wrong. Completing the incorrect addition,

$$\frac{2}{3} + \frac{1}{8} \text{ would be equal to } \frac{2+1}{3+8} = \frac{3}{11},$$

which can't be right. Number lines indicate that the sum should be about 3/4, and 3/11 is nowhere near that.

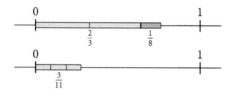

If you find this annoying, I'm right there with you. Mathematical notation is often designed to work conveniently with calculations, so it's natural to feel disgruntled when it doesn't. It's particularly natural to feel disgruntled when the correct calculation procedure is a multistep horror, as it undeniably is for adding fractions.

$$\frac{2}{3} + \frac{1}{8} = \frac{(2 \times 8) + (1 \times 3)}{3 \times 8} = \frac{16 + 3}{24} = \frac{19}{24}.$$

If you learned this procedure without understanding it, and consequently became confused, started getting things wrong, and decided that you didn't like mathematics any more, you have my sympathy, and we'll sort it out. If you did understand it, then before reading on you might want to see whether you can explain why it works. Would your explanation satisfy a sceptic? How about a sceptical 9 year-old?

As you might expect, I'm going to forget the procedure for now and start with the meaning. What are we trying to do? When is it easy and when is it hard? Well, it's easy when we're adding fractions with the same denominator. Adding 3/5 to 4/5 gives 7/5. There's no mystery in that: the 'pieces'—the fifths—are all the same size, so we can just line them up. Three fifths plus four fifths is seven fifths, just as three carrots plus four carrots is seven carrots.

Things are also not too bad if the fractions' denominators work well together. To add 1/2 and 1/4, for instance, it's useful to split the half into two quarters, so that the sum is two quarters plus one quarter, totalling three quarters. To add 1/2 and 1/6, it's useful to split the half into three sixths, so that the sum is three sixths plus one sixth, totalling four sixths.

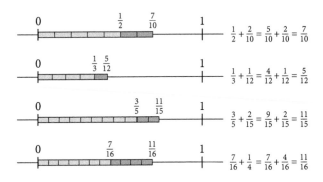

And there is nothing special about these numbers. The same reasoning works whenever one fraction can be rewritten so that the two have a *common denominator*. Once they do, we're back in the carrots situation—we can just line them up.

$$\frac{1}{2} + \frac{2}{10} = \frac{5}{10} + \frac{2}{10} = \frac{7}{10}$$

$$\frac{1}{3} + \frac{1}{12} = \frac{4}{12} + \frac{1}{12} = \frac{5}{12}$$

$$\frac{3}{5} + \frac{2}{15} = \frac{9}{15} + \frac{2}{15} = \frac{11}{15}$$

$$\frac{7}{16} + \frac{1}{4} = \frac{7}{16} + \frac{4}{16} = \frac{11}{16}$$

If you understand that, you understand enough to make sense of the full procedure. The remaining problem is that it's not always possible to rewrite one fraction so that it has the same denominator as the other. For instance, we can't add 1/2 and 1/3 in this way. It's not possible to split the half into thirds—thirds don't 'fit'. What do you think we should do instead? I expect that most readers will either remember or invent the sensible solution: split *both* fractions. What should we split them into? We need something that works for both, and a sensible choice is sixths. Splitting the half gives three sixths, and splitting the third gives two sixths. Adding, therefore, gives five sixths.

Here are some more sums. What splitting would you do for each one, and what are the totals?

$$\frac{1}{2} + \frac{1}{5} \qquad \frac{1}{3} + \frac{1}{4} \qquad \frac{2}{3} + \frac{1}{5} \qquad \frac{2}{3} + \frac{1}{8}$$

And here are some accompanying diagrams. Did you choose the same common denominators?

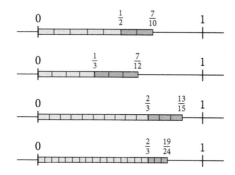

In all of these cases a sensible denominator is the product of the two fractions' denominators. For the last example, for instance, it makes sense to convert to twenty-fourths and calculate

$$\frac{2}{3} + \frac{1}{8} = \frac{16}{24} + \frac{3}{24} = \frac{19}{24}.$$

We could, in fact, write out the multiplication at the first step:

$$\frac{2}{3} + \frac{1}{8} = \frac{2 \times 8}{3 \times 8} + \frac{1 \times 3}{8 \times 3} = \frac{16}{24} + \frac{3}{24} = \frac{19}{24}.$$

Then it is clearer that cross-multiplying as in the standard procedure has the same effect:

$$\frac{2}{3} + \frac{1}{8} = \frac{(2 \times 8) + (1 \times 3)}{3 \times 8} = \frac{16 + 3}{24} = \frac{19}{24}.$$

As usual, though, there's no need to write in a specific way. I marginally prefer the written-out multiplication to the standard cross-multiplying. If you do too, use it. Or pick your own favourite. It's not worth sacrificing understanding to save a tiny bit of ink.

It might sometimes be worth thinking harder about the denominator, though. It's always fine to use the product of the two denominators. But is it always efficient? What if we want to do this sum?

$$\frac{1}{6} + \frac{3}{4}.$$

We could convert to twenty-fourths again. But do we need to? No—twelfths will do. That's because 12 is the lowest *common multiple* of 6 and 4, and we get

$$\frac{1}{6} + \frac{3}{4} = \frac{2}{12} + \frac{9}{12} = \frac{11}{12}.$$

Under what conditions do we need to take the product, and under what conditions can we get away with a lower number? In other words, under what conditions is the least common multiple less than the product? If you have never thought about this, it's worth exploring.

In the meantime, having had some practice, we can confirm the claims about the harmonic series from the beginning of this chapter.

$$1 + \frac{1}{2} \qquad\qquad = \frac{2}{2} + \frac{1}{2} \quad = \frac{3}{2}$$

$$1 + \frac{1}{2} + \frac{1}{3} \qquad\qquad = \frac{3}{2} + \frac{1}{3} \quad = \frac{9}{6} + \frac{2}{6} \quad = \frac{11}{6}$$

$$1 + \frac{1}{2} + \frac{1}{3} + \frac{1}{4} \qquad = \frac{11}{6} + \frac{1}{4} \quad = \frac{22}{12} + \frac{3}{12} \quad = \frac{25}{12}$$

$$1 + \frac{1}{2} + \frac{1}{3} + \frac{1}{4} + \frac{1}{5} \quad = \frac{25}{12} + \frac{1}{5} \quad = \frac{125}{60} + \frac{12}{60} \quad = \frac{137}{60}$$

$$1 + \frac{1}{2} + \frac{1}{3} + \frac{1}{4} + \frac{1}{5} + \frac{1}{6} = \frac{137}{60} + \frac{1}{6} \quad = \frac{137}{60} + \frac{10}{60} \quad = \frac{147}{60}$$

This confirms that my diagram was about right. The fraction $\frac{147}{60}$ is a tiny bit less than $2\frac{1}{2}$, as shown below. If you prefer numerical comparisons, you might like to note that $\frac{150}{60}$ would be exactly $2\frac{1}{2}$ because $60 \times 2\frac{1}{2} = 150$.

How many more fractions would you want to add to confidently predict the total for the infinite sum?

3.4 Adding up lots of numbers

To work toward infinite sums, we will start with finite sums that are longer than those we've looked at so far. Consider, for instance, the sum of the numbers from 1 to 100. No one wants to write this out, so mathematicians again use an ellipsis, writing something like

$$1 + 2 + 3 + \ldots + 98 + 99 + 100.$$

Here the ellipsis means 'and so on *until*'. There's leeway in the specifics, though. In many cases a mathematician could write

$$1 + 2 + \ldots + 100$$

and be confident that readers would know what was intended. People tend to think that mathematics is precise and should be written in only one way. That is partly true—mathematicians care about precision—but what is written comes down to clear communication and thus varies according to audience. And this applies to words as well as symbols. For instance, English speakers use the word 'numbers' with considerable ambiguity. When I wrote 'the sum of the numbers from 1 to 100', I meant 'the sum of the *whole* numbers from 1 to 100'. I definitely didn't mean the sum of the rational numbers (for instance). But I knew that most readers would interpret as I intended, and that anyone who didn't would quickly resolve this by looking at the numerical expression.

In any case, an expression like $1 + 2 + 3 + \ldots + 98 + 99 + 100$ invites us to imagine starting with 1, then adding 2, then adding 3, and so on. Most people conclude that this would keep them busy for a long time. You might have heard, however, that the young Gauss astounded his teacher by quickly announcing the correct answer. The story might well be apocryphal, but it's such a good one that no one much cares—we're all quite taken with the idea of a child genius. Genius stories concern me, though, because they can encourage people to believe that mathematical reasoning requires innate talent and is far removed from what most of us can do. It's not, and I'd like to replace the mystery with the more mundane but empowering sense that even if most of us didn't invent the relevant reasoning, we're perfectly able to follow and appreciate it.

For instance, there are several straightforward ways to calculate the sum of the numbers from 1 to 100. First, imagine writing the numbers from 1 to 50 in a row, then turning around and writing the numbers from 51 to 100 in the other direction underneath.

$$\to \quad 1 \quad 2 \quad 3 \quad 4 \quad 5 \ldots 46 \ 47 \ 48 \ 49 \ 50$$
$$100 \ 99 \ 98 \ 97 \ 96 \ldots 55 \ 54 \ 53 \ 52 \ 51 \leftarrow$$

Now the numbers are in columnnar pairs. What does the first pair add up to? And the second? And the rest? There are 50 pairs and each pair adds up to 101. So the total is $50 \times 101 = 5050$. That didn't take long at all.

Did you notice, too, that this argument uses properties from Chapter 1? The first step relies upon re-ordering the numbers. Instead of

$$1 + 2 + 3 + \ldots + 98 + 99 + 100,$$

we calculate

$$(1 + 100) + (2 + 99) + \ldots + (49 + 52) + (50 + 51).$$

This involves commutativity in a big way: many swaps are required to get the numbers into this new, more convenient order.

The second step, depending on how you tackle it, uses distributivity. If you checked the multiplication, you might have split 101 into $100 + 1$ and reasoned that

$$50 \times 101 = 50 \times (100 + 1) = (50 \times 100) + (50 \times 1) = 5000 + 50 = 5050.$$

In mental calculations, distributivity is very handy.

Also, the whole argument is generalizable in various ways. For instance, can you work out the sum of the numbers from 1 to 200? From 1 to 1000? How about from 11 to 100? From 101 to 1000? And, in the original, do we have to use two rows of 50 numbers? How about four rows of 25 instead?

$$
\begin{array}{l}
\rightarrow \quad 1 \quad 2 \quad 3 \quad 4 \quad 5 \ldots 21\ 22\ 23\ 24\ 25 \\
 \quad 50\ 49\ 48\ 47\ 46 \ldots 30\ 29\ 28\ 27\ 26 \leftarrow \\
\rightarrow \quad 51\ 52\ 53\ 54\ 55 \ldots 71\ 72\ 73\ 74\ 75 \\
 \quad 100\ 99\ 98\ 97\ 96 \ldots 80\ 79\ 78\ 77\ 76 \leftarrow
\end{array}
$$

Does that still work? If so, what multiplication do we end up doing? If not, what goes wrong? And what other variations are possible? Could we add the numbers from 1 to 300 by arranging them in three rows? As ever, the original argument is cute but we can learn more by generalizing.

An alternative way to think about the sum is to visualize. An image for the sum from 1 to 100 would occupy too much space, so I'll do 1 to 7 and invite you to see that as generic. Here are some dots representing the sum $1 + 2 + 3 + 4 + 5 + 6 + 7$.

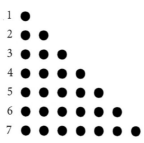

And here's the clever bit: adjoin another copy of this sum the other way up.

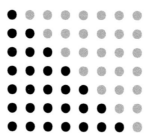

Together these make a rectangle, and in a rectangle it's easy to find the total number of dots. Here the total is $7 \times 8 = 56$ (if you thought it would be $7 \times 7 = 49$, look carefully—this isn't quite a square). The original sum $1 + 2 + 3 + 4 + 5 + 6 + 7$ must be half of this, so it is $56/2 = 28$.

Now, what would we calculate to add the numbers from 1 to 100? In that case there would be 100 rows, so the rectangle would have $100 \times 101 = 10100$ dots. For the sum we'd want half of them, once again giving the total $10100/2 = 5050$.

To generalize fully, mathematicians might ask about the sum

$$1 + 2 + 3 + \ldots + n,$$

where n could be any natural number. The following is a formula for that sum. Can you see how this would relate to a triangle with n rows?

$$1 + 2 + 3 + \ldots + n = \frac{n(n+1)}{2}.$$

As you will know if you read Chapter 1, I really like visual arguments. I find them wonderfully compelling. Not everyone does, though, and opinions tend to vary with mathematical experience. Nonexperts often find diagrams completely convincing: they see them as generic and believe that whatever is claimed will always work. People with more mathematical training, though, often mistrust them. At some point they are told that 'a picture isn't a proof' because it shows only one case, and they learn that mathematicians expect written, deductive arguments, probably containing algebra. They're right in a sense—mathematicians do expect students to write careful algebraic arguments. That's only partly due to the one-case problem, however. Mathematicians also value theory building, and algebraic arguments often show how a theory fits together.

The next section will consider some algebraic arguments. If you like visualizing, though, you might like to know that there are entire books of *proofs without words* and that you can find loads on the internet. Here is another, this time about adding odd numbers. What is n in the diagram, and how does the diagram correspond to the formula? Does the diagram work as a proof without words for you?

$$1 + 3 + 5 + \ldots + (2n - 1) = n^2$$

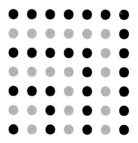

If not, some extra labels might help. In the following, the left diagram highlights the grouping of dots into inverted L-shapes, and shows how many dots are in each. This diagram thus represents $1+3+5+7+9+11+13$ and, relating this to the general sum $1+3+5+\ldots+(2n-1)$, we should have $2n-1=13$, so $n=7$. The sum should thus be 7^2, which it is—the square of dots has seven dots along each edge. Alternatively, the right diagram focuses attention on the nth L-shape of an $n \times n$ square. This L-shape comprises a horizontal set of n dots and a vertical set of n dots, but these overlap in the dot in the top right corner. So the nth L-shape has $2n-1$ dots, and n^2 is equal to the sum of odd numbers up to and including the odd number $2n-1$.

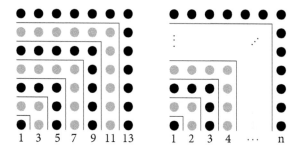

3.5 Adding up lots of odd numbers

If you're not a diagram fan—if you prefer orderly logical arguments— then you might like an alternative way of reasoning about the sum

$$1 + 3 + 5 + \ldots + (2n - 1) = n^2.$$

One thing to note is that although this looks like a single equation, it captures infinitely many sums, one for each natural number n.

For $n = 1$: 1 $= 1^2$
For $n = 2$: $1 + 3$ $= 2^2$
For $n = 3$: $1 + 3 + 5$ $= 3^2$
For $n = 4$: $1 + 3 + 5 + 7 = 4^2$, and so on.

Make sure you believe that the list uses n's correctly. For $n = 4$, for instance, $2n - 1 = 7$, so the sum on the left-hand side should stop at 7 (which it does) and the number on the right-hand side should be 4^2 (which it is, and which checks out: $1 + 3 + 5 + 7 = 16 = 4^2$). Check the others and, if you find the first one peculiar, note that you're bound to because it's a degenerate case: it involves 'adding up' just one thing, which doesn't feel like adding up at all. But the formula nevertheless works in the same way.

Another thing to consider is the expression $2n - 1$ as the final number in the sum. This always gives an odd number: whatever n is, doubling it gives an even number, so doubling it and subtracting 1 gives an odd number. Note also that whatever n is, the next odd number after $2n - 1$ will be $2n + 1$. For $n = 4$, for instance, $2n - 1 = 7$ and $2n + 1 = 9$. We could think about this on a number line. If that doesn't convince you that it always works, try it with other values of n.

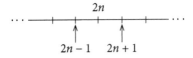

Now, we could keep on forever writing out the equation for different values of n, in each case checking that the sum is the square number as claimed. But that wouldn't be very interesting, and it wouldn't provide much insight into why all these equations should be valid. One way to improve on this is to use an *inductive argument*, in which instead of considering each equation separately, we consider how each one is related to the next. For example, suppose we know that the equation is valid for $n = 8$. For $n = 8$, $2n - 1 = 15$ and $n^2 = 64$, so this means that

$$1 + 3 + 5 + 7 + 9 + 11 + 13 + 15 = 64.$$

You can check the addition if you like, but that's not the point here. The point is, what does knowing about the $n = 8$ case tell us about the $n = 9$ case? We could start from scratch for $n = 9$, but we don't need to. All we need do is take the $n = 8$ case and add one more odd number, ensuring that we do the same to both sides of the equation:

$$(1 + 3 + 5 + 7 + 9 + 11 + 13 + 15) + 17 = 64 + 17 = 81.$$

That gives $81 = 9^2$ as expected, which is good as far as it goes. But again it doesn't provide insight into *why* the new sum must be the next square number. To acquire that insight, we need to work in a more general way.

Suppose we know that the equation is valid for a nonspecific natural number, say $n = k$. That means we know that

$$1 + 3 + 5 + \ldots + (2k - 1) = k^2.$$

What does that imply about the 'next' case, where $n = k + 1$? To find out, we can add the next odd number, $2k + 1$, to both sides:

$$1 + 3 + 5 + \ldots + (2k - 1) + (2k + 1) = k^2 + (2k + 1).$$

Now, that doesn't look finished, because the thing on the right-hand side is supposed to be $(k + 1)^2$. But $(k + 1)^2 = k^2 + 2k + 1$ (review Section 1.5 if you're not sure why). So we do indeed have what we expect, that

$$1 + 3 + 5 + \ldots + (2k - 1) + (2k + 1) = (k + 1)^2.$$

Reviewing this argument's logical structure, we've established that *if* it is true that

$$1 + 3 + 5 + \ldots + (2k - 1) = k^2,$$

then it is true that

$$1 + 3 + 5 + \ldots + (2k - 1) + (2k + 1) = (k + 1)^2.$$

In other words, we've proved that if the equation holds for $n = k$ then it also holds for $n = k + 1$. This might not seem sufficient—you might ask, what if it *isn't* true for $n = k$? But that turns out to be irrelevant, because we can glue together this *if… then…* reasoning with what we know about the 'early' cases. We know the equation holds for $n = 4$, for instance—we checked that by calculating. But that means that the equation must hold for $n = 5$. And that means that it must hold for $n = 6$, and that means that it must hold for $n = 7$, and so on forever. So the equation holds for every value of n.

I think this is a great structure for an argument. We want to prove infinitely many related claims that can be put in a numbered list. We don't want to do infinitely many calculations, so instead we do just the first one (or some early, easy one) then prove that if any claim in the list is valid, so is the next one. This proves that all the claims are true. Elegant, no? Such an argument is known as a *proof by mathematical induction*. And you might find a proof by induction easier to follow if it's written more concisely. The version that follows uses $n = 1$ as the starting case but is otherwise the same as the argument described above. When reading concise mathematics, remember what I said in the introduction: you will probably need to read back and forth, checking each link to work out how the whole thing fits together.

Claim: For every natural number n, $1 + 3 + 5 + \ldots + (2n - 1) = n^2$.

Proof: $1 = 1^2$, so the claim is true for $n = 1$.

Suppose that the claim is true for $n = k$.

This means that $1 + 3 + 5 + \ldots + (2k - 1) = k^2$.

Adding the next odd number, $2k + 1$, to both sides gives

$$1 + 3 + 5 + \ldots + (2k - 1) + (2k + 1) = k^2 + (2k + 1)$$
$$= k^2 + 2k + 1$$
$$= (k + 1)^2.$$

So we have established that if the claim is true for $n = k$, then it is also true for $n = k + 1$.

So, because the claim is true for $n = 1$, it must be true for every natural number n.

Do you like the concise version? Do you understand it? If not, don't give up too quickly—maybe read this section again then have another go. If you do understand it, could you explain it to someone else, and would you have understood it if you hadn't read the wordy version? I think this is an interesting question about presenting mathematics. On the one hand, I can write as I would talk, explaining in detail with lots of examples that are hopefully illuminating but not logically necessary.

In doing this I'm aiming for accessibility, but the explanation gets long. On the other hand, I can write as I would for a more mathematically experienced audience, with a concise claim and proof. This compresses the argument, which makes its overall structure more salient. But it places a higher burden on the reader to follow the thinking, working out why each step works and how they all fit together. To understand how mathematicians think, I think it helps to appreciate attempts to balance concision and clarity.

To conclude this section, I'd like to return to the philosophical point about representations. Comparing an algebraic argument with a diagrammatic 'proof' does highlight obvious differences. An algebraic argument makes more steps explicit; it lays out a linear chain of reasoning, whereas a diagram presents all the information at once so that the reader must work out what to think about in what order. Partly because of this, some people find it easier to be confident that algebraic arguments 'work for everything'. But algebraic arguments still require interpretation: we must believe that the n in the sum and the formula really can represent any natural number, checking at every step if necessary. People who get used to algebra become confident about standard manipulations, so they can offload the believing-it-always-works problem to those manipulations. But the believing-it issue remains, it's just pushed into the background. So visual and algebraic arguments are less different in this respect than they might appear.

3.6 Powers of 2

If you enjoyed the inductive argument, there is another to look forward to in this section. If you didn't—if you got a bit lost—then in the next section you'll find a numerical argument that is, in my view, even more elegant. Both are about adding up powers of 2, so we'll start with information on those.

In some cases, the idea of powers is fairly straightforward:

$$2^2 = 2 \times 2$$
$$2^3 = 2 \times 2 \times 2$$
$$2^4 = 2 \times 2 \times 2 \times 2$$
$$2^5 = 2 \times 2 \times 2 \times 2 \times 2, \text{ and so on.}$$

The number 2-to-the-power-4, for instance, is '2 multiplied by itself four times' (I'm not keen on this phrasing, but it is used a lot). Mathematicians sometimes capture this using 'underbrace' notation:

$$2^4 = \underbrace{2 \times \ldots \times 2}_{4 \text{ times}} \qquad 2^n = \underbrace{2 \times \ldots \times 2}_{n \text{ times}}$$

It will be useful to observe that $2 \times 2^n = 2^{n+1}$, because

$$2 \times 2^2 = 2 \times (2 \times 2) = 2^3, \text{ and}$$
$$2 \times 2^3 = 2 \times (2 \times 2 \times 2) = 2^4, \text{ and so on.}$$

Also, $2^3 \times 2^5 = 2^8$ because, when multiplying 2^3 by 2^5, we end up with eight 2s multiplied together:

$$2^3 \times 2^5 = \underbrace{(2 \times 2 \times 2)}_{3 \text{ times}} \times \underbrace{(2 \times 2 \times 2 \times 2 \times 2)}_{5 \text{ times}} = 2^8.$$

The result can be written as $2^3 2^5 = 2^{3+5}$, which is a special case of the general rule that $x^m x^n = x^{m+n}$. If you tend to get confused by this, that's because you're normal. Remember how people often mistakenly say that $(a+b)^2 = a^2 + b^2$? This is similar. The desire for tidy notation might make us think that the rule should be $x^m x^n = x^{mn}$. But it isn't. To remember the correct version, I tend to think about a specific case like $2^3 2^5 = 2^{3+5}$, or to use underbrace notation for the general case:

$$x^m x^n = \underbrace{(x \times \ldots \times x)}_{m \text{ times}} \times \underbrace{(x \times \ldots \times x)}_{n \text{ times}} = x^{m+n}.$$

The other thing that often baffles people is what happens when a power is zero or negative. How do we 'multiply 2 by itself' zero times or minus 3 times? Do you know or can you remember? I find that new undergraduate mathematics students do know: they can confidently say that 'anything to the power zero is 1', for instance. But when I ask why, they often have no idea. In fact, they look a little confused by the question, and these thoughts pass across their faces:

> 'Well, because my teacher said so.'
> 'Oh, I guess that's not a very mathematical reason.'
> 'Um... and I don't know a better one.'
> 'Oh dear, that's embarrassing.'

But there's no need for embarrassment—this is easier to fix than you might think. My favourite approach is to think about powers lined up as follows. Note that moving one step to the right involves multiplying by 2.

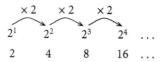

What about moving to the left? Moving left involves dividing by 2. But that means that if we keep going to the left, we discover the values of 2^0 and 2^{-1} and so on.

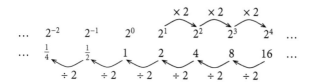

Isn't that nice? When I show this to previously embarrassed undergraduates, their faces go all open as if to say, 'I had no idea it was that simple!' I enjoy this a lot—it's nice watching people gain insight. From a more sophisticated perspective, this argument is a way to think about consistency within number systems: it says that powers of 2 will be defined for non-obvious cases by extending existing relationships. There are more formal ways to think about such consistency, and we'll revisit this idea at the end of Chapter 5.

For now, let's return to adding up. Here is a general claim about adding powers of 2:

$$1 + 2 + 4 + 8 + \ldots + 2^{n-1} = 2^n - 1.$$

As before, this is really infinitely many claims.

For $n = 1$: 1 $= 2^1 - 1$
For $n = 2$: $1 + 2$ $= 2^2 - 1$
For $n = 3$: $1 + 2 + 4$ $= 2^3 - 1$
For $n = 4$: $1 + 2 + 4 + 8 = 2^4 - 1$, and so on.

And as before, these ones are straightforward to check. For $n = 4$, for instance, $n - 1 = 3$, so the sum on the left-hand side should stop at 2^3 (which it does) and the number on the right should be $2^4 - 1$ (which it is, and which checks out: $1 + 2 + 4 + 8 = 15 = 16 - 1 = 2^4 - 1$). But, as before, we wouldn't want to keep checking, and we can construct an inductive argument to prove the whole lot. This time, I'll straightaway write a concise claim and proof. Even if you weren't sure about the previous inductive argument, try reading this one—sometimes more exposure does the trick.

Claim: For every natural number n, $1 + 2 + 4 + 8 + \ldots + 2^{n-1} = 2^n - 1$.

Proof: $1 = 2^1 - 1$, so the claim is true for $n = 1$.

Suppose that the claim is true for $n = k$.

This means that $1 + 2 + 4 + 8 + \ldots + 2^{k-1} = 2^k - 1$.

Adding 2^k to both sides gives

$$
\begin{aligned}
1 + 2 + 4 + 8 + \ldots + 2^{k-1} + 2^k &= (2^k - 1) + 2^k \\
&= 2^k + 2^k - 1 \\
&= 2 \times 2^k - 1 \\
&= 2^{k+1} - 1.
\end{aligned}
$$

So we have established that if the claim is true for $n = k$, then it is also true for $n = k + 1$.

So, because the claim is true for $n = 1$, it must be true for every natural number n.

If you found that a struggle, try some specific examples: what would the claim say for $n = 8$, and how would that relate to the claim for $n = 9$? If you found it easy, ask yourself whether something similar would work for powers of 3. If you're really getting into inductive arguments, try constructing one to prove that for every natural number n,

$$
1^2 + 2^2 + \ldots + n^2 = \frac{n(n + 1)(2n + 1)}{6}.
$$

If none of that sounds appealing and you wish I would give up the algebra and go back to numbers, read on.

3.7 Adding up powers

To add powers of 2, another approach is more numerical and particularly elegant. Suppose we want to add up the first ten powers of 2 (starting at $2^0 = 1$). It helps to give the sum a name, and we'll call it S.

$$S = 1 + 2 + 4 + 8 + 16 + 32 + 64 + 128 + 256 + 512.$$

Now double every number in the sum, so that the new sum is $2S$.

$$2S = 2 + 4 + 8 + 16 + 32 + 64 + 128 + 256 + 512 + 1024.$$

Next, change nothing but line up the two sums so that their similarities are obvious.

$$2S = \phantom{1 + {}} 2 + 4 + 8 + 16 + 32 + 64 + 128 + 256 + 512 + 1024$$
$$S = 1 + 2 + 4 + 8 + 16 + 32 + 64 + 128 + 256 + 512$$

Finally, subtract the bottom row from the top. On the left-hand side, this gives $2S - S = S$. On the right-hand side, most of the addends 'cancel out', leaving $1024 - 1$. So the sum is $S = 1023$.

This approach works because this sum is *geometric*—the ratio of each term to its predecessor is the same. Here the ratio is 2, and we could make that more obvious and write the argument concisely as follows:

Let $\quad S = 2^0 + 2^1 + 2^2 + 2^3 + 2^4 + 2^5 + 2^6 + 2^7 + 2^8 + 2^9.$

Then $\quad 2S = \phantom{2^0 + {}} 2^1 + 2^2 + 2^3 + 2^4 + 2^5 + 2^6 + 2^7 + 2^8 + 2^9 + 2^{10}.$

So $\quad 2S - S = 2^{10} - 2^0,$

$\Rightarrow \quad S = 1023.$

There is nothing special about the common ratio 2—similar arguments work for other ratios. And we don't have to start at the number 1. For instance, we could add up a shorter sum of powers of 3.

$$\text{Let} \qquad S = 3^5 + 3^6 + 3^7 + 3^8 + 3^9.$$
$$\text{Then} \qquad 3S = \qquad \; 3^6 + 3^7 + 3^8 + 3^9 + 3^{10}.$$
$$\text{So} \quad 3S - S = 3^{10} - 3^5$$
$$\Rightarrow \qquad 2S = 59\,049 - 243.$$
$$\Rightarrow \qquad 2S = 58\,806$$
$$\Rightarrow \qquad S = 29\,403.$$

I used my calculator for the last few lines—I'm no mental arithmetic genius. That being the case, do you think this method is worth it for sums this short? Would you, instead, use a calculator from the beginning? I tend to resist calculator temptation because working intelligently on paper feels more elegant, and because I'm liable to press the wrong buttons. What I tell my students is that calculators are fast but stupid—they will give you the answer you asked for, whether or not it is the one you want. It can be a good idea to minimize calculator work by first doing some thoughtful manipulations. But I digress.

The astute reader might notice that in arguments like this, the common ratio *does not have to be a whole number*. Suppose that instead of powers of 2, we add powers of $\frac{1}{2}$. Let's take ten terms again and consider the corresponding argument,

$$\text{Let} \qquad S = 1 + \frac{1}{2} + \frac{1}{4} + \frac{1}{8} + \frac{1}{16} + \frac{1}{32} + \frac{1}{64} + \frac{1}{128} + \frac{1}{256} + \frac{1}{512}.$$
$$\text{Then} \quad \frac{1}{2}S = \quad \frac{1}{2} + \frac{1}{4} + \frac{1}{8} + \frac{1}{16} + \frac{1}{32} + \frac{1}{64} + \frac{1}{128} + \frac{1}{256} + \frac{1}{512} + \frac{1}{1024}.$$
$$\text{So} \quad S - \frac{1}{2}S = 1 - \frac{1}{1024}$$
$$\Rightarrow \qquad \frac{1}{2}S = 1 - \frac{1}{1024}$$
$$\Rightarrow \qquad S = 2 - \frac{2}{1024}$$
$$\Rightarrow \qquad S = 2 - \frac{1}{512}.$$

How is this argument like the previous ones and how does it differ? And would you have done one more calculation at the end to give a single fraction rather than $2 - \frac{1}{512}$? You could, but I think it's more informative to leave it as $2 - \frac{1}{512}$ because that makes clear that the sum is a tiny bit less than 2. And that sets us up to return to the infinite sum that began this chapter.

3.8 The geometric series $1 + \frac{1}{2} + \frac{1}{4} + \frac{1}{8} + \frac{1}{16} + \frac{1}{32} + \ldots$

This chapter opened with the geometric series $1 + \frac{1}{2} + \frac{1}{4} + \frac{1}{8} + \frac{1}{16} + \frac{1}{32} \ldots$ and the problem of reaching the door. I presented the following diagram and invited you to agree that the fractions add up to 2. I also acknowledged that you might be unhappy about that because you might think that the sum is a tiny bit less than 2.

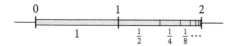

If you do think that, you need to know that this is partly an issue of notation. The ellipsis in $1 + \frac{1}{2} + \frac{1}{4} + \frac{1}{8} + \frac{1}{16} + \frac{1}{32} + \ldots$ invites many people to think of adding as a process that happens in time. You imagine starting with 1, then adding $\frac{1}{2}$, then adding $\frac{1}{4}$, and so on. Each of these finite sums is, indeed, a bit less than 2.

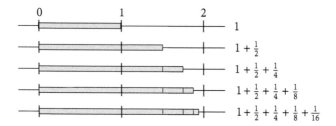

But mathematicians don't interpret the ellipsis that way. When they write '…' without any 'ending' number, they mean 'and so on *forever*'. That is, they think of $1 + \frac{1}{2} + \frac{1}{4} + \frac{1}{8} + \frac{1}{16} + \frac{1}{32} + \ldots$ as representing not a process that happens in time, but an entire sum in which all the terms—all infinitely many of them—are already there.

This might seem weird, and there is some inherent ambiguity because mathematical notations often do capture both process and object. For a small child, for instance, the sum $5 + 3$ might be very much a process that happens in time. This process might be quite long: counting out five cubes then counting out another three then counting them all together. Or it might be shorter: starting at 5 and counting on ('6, 7, 8'). But more mathematically experienced people can interpret the sum $5 + 3$ not as a process, but as one way to write a number that can also be written in other ways: $3 + 5$ or 8 or 2^3, for instance.

The same applies to $1 + \frac{1}{2} + \frac{1}{4} + \frac{1}{8} + \frac{1}{16} + \frac{1}{32} + \ldots$. This might make you imagine a process that happens in time. And, because the sum is infinite, you might feel that it has to be imagined this way. But it doesn't. I bet that if you try, you can imagine that the whole sum is already written down.

It's worth thinking about this series in a couple of ways, because different people find different things more natural or convincing. One way is simply to extend the previous type of argument. In this case there is no 'ending' number and, because the list is infinite, subtracting cancels every number except 1.

$$\text{Let} \qquad S = 1 + \frac{1}{2} + \frac{1}{4} + \frac{1}{8} + \frac{1}{16} + \frac{1}{32} + \frac{1}{64} + \frac{1}{128} + \frac{1}{256} + \frac{1}{512} + \ldots.$$

$$\text{Then} \quad \frac{1}{2}S = \qquad \frac{1}{2} + \frac{1}{4} + \frac{1}{8} + \frac{1}{16} + \frac{1}{32} + \frac{1}{64} + \frac{1}{128} + \frac{1}{256} + \frac{1}{512} + \ldots.$$

$$\text{So} \quad S - \frac{1}{2}S = 1$$

$$\Rightarrow \qquad \frac{1}{2}S = 1$$

$$\Rightarrow \qquad S = 2$$

There is nothing fishy about that. But, if the lack of an ending number makes you uncomfortable, you might prefer a more formal argument that starts with finitely many terms then considers what happens *in the limit*. The finite case works in the familiar way.

Let
$$S = 1 + \frac{1}{2} + \left(\frac{1}{2}\right)^2 + \left(\frac{1}{2}\right)^3 + \ldots + \left(\frac{1}{2}\right)^n.$$

Then
$$\frac{1}{2}S = \frac{1}{2} + \left(\frac{1}{2}\right)^2 + \left(\frac{1}{2}\right)^3 + \ldots + \left(\frac{1}{2}\right)^n + \left(\frac{1}{2}\right)^{n+1}.$$

So
$$S - \frac{1}{2}S = 1 - \left(\frac{1}{2}\right)^{n+1}$$

$$\Rightarrow \quad \frac{1}{2}S = 1 - \left(\frac{1}{2}\right)^{n+1}$$

$$\Rightarrow \quad S = 2 - 2\left(\frac{1}{2}\right)^{n+1}.$$

$$\Rightarrow \quad S = 2 - \left(\frac{1}{2}\right)^n.$$

So the sum of these terms is $S = 2 - \left(\frac{1}{2}\right)^n$. As n gets larger, $\left(\frac{1}{2}\right)^n$ gets smaller. In the limit, the difference between S and 2 is smaller than $\frac{1}{2}$, and smaller than $\frac{1}{4}$, and smaller than $\left(\frac{1}{2}\right)^n$ for *every possible* n. So the difference must be 0, meaning that $S = 2$.

If that leaves you feeling weird, it's likely because you feel that the difference should be *infinitesimal*: smaller than every positive number yet bigger than 0. This idea is intuitively appealing and there are ways to formalize it, but they are not straightforward. If infinitesimals exist, for instance, what happens if we add together two of them, or a hundred, or infinitely many? Does any of these give a 'normal' number? Fortunately, the mathematics of the number line works perfectly under the 'standard' interpretation in which a nonnegative number that is smaller than every other number must be 0. So we'll stick to that.

My favourite thing about geometric series is that some can be captured in truly lovely diagrams. For instance, an argument similar to the one we've been discussing but using powers of $\frac{1}{4}$ proves that

$$\frac{1}{4} + \frac{1}{16} + \frac{1}{64} + \frac{1}{256} + \frac{1}{1024} + \ldots = \frac{1}{3}.$$

Here is the start of an argument of the 'no ending number' style. Can you finish it?

Let
$$S = \frac{1}{4} + \frac{1}{16} + \frac{1}{64} + \frac{1}{256} + \frac{1}{1024} + \dots.$$

Then
$$\frac{1}{4}S = \quad\ \frac{1}{16} + \frac{1}{64} + \frac{1}{256} + \frac{1}{1024} + \dots.$$

. . .

And can you 'see' that $\frac{1}{4} + \frac{1}{16} + \frac{1}{64} + \frac{1}{256} + \frac{1}{1024} + \dots = \frac{1}{3}$ by thinking about the black and white square below? If the area of the whole square is 1, then the area of the biggest black square is $\frac{1}{4}$, the area of the next biggest is $\frac{1}{4} \times \frac{1}{4} = \frac{1}{16}$, and so on. What proportion of the whole square is black? If you think squares are a bit clunky and triangles are more elegant (as I do) this can be done with triangles too.

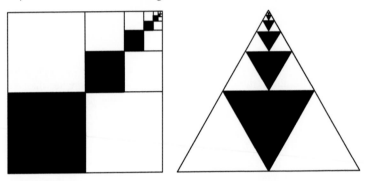

3.9 The harmonic series $1+\frac{1}{2}+\frac{1}{3}+\frac{1}{4}+\frac{1}{5}+\frac{1}{6}+\dots$

The other series that began this chapter was the harmonic series

$$1 + \frac{1}{2} + \frac{1}{3} + \frac{1}{4} + \frac{1}{5} + \frac{1}{6} + \dots.$$

Again this is an infinite sum—the ellipsis means that all the terms are already there. And it has much in common with the geometric series we've been working with: it starts at 1 and adds progressively smaller fractions. I've asked a couple of times what you think the sum might be. What do you think now?

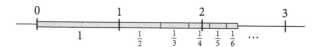

Most people expect the total to be quite small, maybe around 4 or 5, and certainly less than 10. But this turns out to be very, very wrong. In fact, the total is infinite. It's bigger than any number you could name. You might be sceptical about that, because it's very counterintuitive. Many people think, 'But look, what we're adding gets smaller and smaller—eventually we'll be adding millionths of millionths, and those will make hardly any difference'. That's entirely reasonable and it's definitely what I thought when I first encountered this series. But I was wrong. And that's why I think series are so interesting now. Bear with me and I'll explain how this works, then discuss counterintuitive results in general.

The first term of the harmonic series is 1. The second is $\frac{1}{2}$. After that, we'll do some clever thinking. The third term and the fourth together add up to something greater than $\frac{1}{2}$. Because $\frac{1}{3} > \frac{1}{4}$, we get

$$\frac{1}{3} + \frac{1}{4} > \frac{1}{4} + \frac{1}{4} = \frac{1}{2}.$$

The exact value of $\frac{1}{3} + \frac{1}{4}$ isn't necessary for this argument. But here is another use for underbrace notation, to group together terms of the series.

$$1 + \frac{1}{2} + \underbrace{\frac{1}{3} + \frac{1}{4}}_{>\frac{1}{2}} + \frac{1}{5} + \frac{1}{6} + \frac{1}{7} + \frac{1}{8} + \frac{1}{9} + \frac{1}{10} + \frac{1}{11} + \frac{1}{12} + \frac{1}{13} + \frac{1}{14} + \frac{1}{15} + \frac{1}{16} + \cdots$$

Similarly, each of the next four terms is greater than or equal to $\frac{1}{8}$. So the four together again add up to something greater than $\frac{1}{2}$.

$$\frac{1}{5} + \frac{1}{6} + \frac{1}{7} + \frac{1}{8} > \frac{1}{8} + \frac{1}{8} + \frac{1}{8} + \frac{1}{8} = \frac{1}{2}.$$

And this can be represented in underbrace notation.

$$1 + \frac{1}{2} + \underbrace{\frac{1}{3} + \frac{1}{4}}_{>\frac{1}{2}} + \underbrace{\frac{1}{5} + \frac{1}{6} + \frac{1}{7} + \frac{1}{8}}_{>\frac{1}{2}} + \frac{1}{9} + \frac{1}{10} + \frac{1}{11} + \frac{1}{12} + \frac{1}{13} + \frac{1}{14} + \frac{1}{15} + \frac{1}{16} + \cdots$$

How many terms do you think we should consider next? Each of the next eight terms is greater than or equal to $\frac{1}{16}$, so together they add up to more than $\frac{1}{2}$.

$$1 + \frac{1}{2} + \underbrace{\frac{1}{3} + \frac{1}{4}}_{>\frac{1}{2}} + \underbrace{\frac{1}{5} + \frac{1}{6} + \frac{1}{7} + \frac{1}{8}}_{>\frac{1}{2}} + \underbrace{\frac{1}{9} + \frac{1}{10} + \frac{1}{11} + \frac{1}{12} + \frac{1}{13} + \frac{1}{14} + \frac{1}{15} + \frac{1}{16}}_{>\frac{1}{2}} + \ldots$$

And the next 16 terms add another half, and the next 32 add another, and so on. Eventually we need many thousands of terms to add another half. But that's not a problem because there are infinitely many—we never run out. So the infinite sum is bigger than any number of halves added together. It's bigger than 100 halves, and bigger than 1000 halves, and bigger than 1 000 000 halves, and so on. So the total must be infinite.

When I first saw this, I was astonished by the result and captivated by the argument, so much so that I remember exactly where I was: with my supervisor Jean Flower in the undergraduate common room in the Mathematics Institute at the University of Warwick. I was astonished because my intuition was typical: I believed that because the addends kept getting smaller, eventually the sum would 'settle down'. I was captivated by the argument because it is so simple. It convinced me beyond doubt that the total is infinite and that my intuition was completely wrong.

I found that thrilling, and I still do, for several reasons. First, I really like arguments that I can grasp holistically, that I can understand as single ideas rather than multiple steps. That's why I like diagrams and proofs without words. This argument has that quality for me, though it might not for you—I've thought about it a lot so I've compressed it in my mind. If, to you, it still seems long, then you might like to read it a few more times or explain it to someone else—see if that brings about some compression.

Second, I really like mathematical surprises. Some people don't, because surprises make them nervous. Perhaps they suspect that they're being duped, or that whole swathes of their understanding might be faulty. I think the reason I don't feel like that is that I'm confident about the mathematics that I understand. And you should feel confident too. If you're understanding this book—maybe even enjoying it—that's because mathematics is a humanly graspable network of ideas with enormous internal coherence. If you understand big chunks of that network—if you are developing a sense of how it fits together—then the chunks that you've mastered will not fall apart. If you can be confident of that, you might find that you can look at surprises not as threats but as fascinating curiosities. To me, a surprise indicates not that my current understanding is faulty, but that a smallish-looking area of the network that I've never really thought about must be twisted in on itself in a way that accommodates a lot of weirdness. And that makes it worth a closer look.

3.10 Convergence and divergence

So what is going on here? How can we untangle the weirdness so that the harmonic series no longer seems so counterintuitive? Well, one issue is notation. We write series so that most terms are squashed into an 'and the rest' blob denoted by the ellipsis. Attention is focused on the first few terms, which makes the harmonic and geometric series look pretty similar.

$$1 + \frac{1}{2} + \frac{1}{3} + \frac{1}{4} + \ldots$$

$$1 + \frac{1}{2} + \frac{1}{4} + \frac{1}{8} + \ldots$$

But really they're not. Extending even a little shows that the terms of the geometric series shrink much, much faster.

$$1 + \frac{1}{2} + \frac{1}{3} + \frac{1}{4} + \frac{1}{5} + \frac{1}{6} + \frac{1}{7} + \frac{1}{8} + \frac{1}{9} + \frac{1}{10} + \ldots$$

$$1 + \frac{1}{2} + \frac{1}{4} + \frac{1}{8} + \frac{1}{16} + \frac{1}{32} + \frac{1}{64} + \frac{1}{128} + \frac{1}{256} + \frac{1}{512} + \ldots$$

So an informal way to think about the counterintuitive result is that although the terms of the harmonic series get smaller, they *don't get smaller fast enough*. Of course, that raises a question: what constitutes fast enough? A student asked me this recently. She was very intelligent and really quite worried, because this made it seem to her that the results were arbitrary. If observing that the terms get really small was not enough to decide that the sum of a series is finite, how would we know?

In mathematical terms this is a question about *convergence*. We ask which series *converge* to some finite number, and which *diverge* to infinity. Answers exist for many different series, and in this final section I'll cover just enough to give you a taste of the theory.

First, all geometric series with common ratio less than 1 converge, by applications of the argument in Section 3.7. That is, if $0 < r < 1$, we always have a valid argument like this.

$$\text{Let} \qquad S = 1 + r + r^2 + r^3 + r^4 + r^5 + \ldots.$$

$$\text{Then} \qquad rS = \quad\;\; r + r^2 + r^3 + r^4 + r^5 + \ldots.$$

$$\text{So} \qquad S - rS = 1$$

$$\Rightarrow \qquad (1 - r)S = 1$$

$$\Rightarrow \qquad\quad S = \frac{1}{1 - r}.$$

You might like to replace r with some appropriate numbers and think about what the formula says. And you might like to know that this works for r with $-1 < r \leq 0$, too. If you've found these arguments fairly straightforward, work through one with $r = -\frac{1}{2}$, then ask yourself what happens for $r \geq 1$ or $r \leq -1$.

Second, the harmonic series invites comparison with similar series. For instance, consider

$$\frac{1}{1^2} + \frac{1}{2^2} + \frac{1}{3^2} + \frac{1}{4^2} + \frac{1}{5^2} + \frac{1}{6^2} + \ldots.$$

What do you think happens to this series? Does it converge to a finite number, or is the total infinite? I think this is not obvious. On the one hand, this is 'like' the harmonic series except that the denominators are of the form n^2 instead of n. So maybe it diverges to infinity. On the other hand, its terms get smaller considerably faster, so maybe it behaves more like a geometric series. What do you reckon? If you find that you really don't know, that's good. It means that you've understood well enough to recognize that both outcomes are plausible.

To get some traction, a sensible strategy is to compare with known series, perhaps using the following lists. Does this help?

$$\text{harmonic: } 1 + \frac{1}{2} + \frac{1}{3} + \frac{1}{4} + \frac{1}{5} + \frac{1}{6} + \frac{1}{7} + \frac{1}{8} + \frac{1}{9} + \frac{1}{10} + \ldots$$

$$\text{new: } 1 + \frac{1}{4} + \frac{1}{9} + \frac{1}{16} + \frac{1}{25} + \frac{1}{36} + \frac{1}{49} + \frac{1}{64} + \frac{1}{81} + \frac{1}{100} + \ldots$$

$$\text{geometric: } 1 + \frac{1}{2} + \frac{1}{4} + \frac{1}{8} + \frac{1}{16} + \frac{1}{32} + \frac{1}{64} + \frac{1}{128} + \frac{1}{256} + \frac{1}{512} + \ldots$$

If the terms of the new series were bigger than the corresponding terms of the harmonic series, we could say that the new one would diverge. But they are in fact smaller, and knowing that they are smaller doesn't help because we don't know where the cut-off for 'gets smaller fast enough' is. Unfortunately, comparing with the geometric series doesn't help either. The first six terms of the new series are less than or equal to their counterparts in the geometric one with common ratio $\frac{1}{2}$, which looks promising. After that, though, they're bigger, so we still don't know whether they get smaller fast enough. If you have good grasp of what's going on here, it might occur to you to try comparing with a different geometric series with a bigger common ratio, like $\frac{3}{4}$ or $\frac{9}{10}$. Write out enough terms, though (maybe in a spreadsheet if you don't feel like doing the calculations), and you'll find that doesn't work either: eventually, the terms of the geometric series will be smaller.

Nevertheless, it turns out that the new series does converge, and we will establish that and round off this chapter by pulling together three of its key ideas: comparing fractions, adding fractions, and induction.

We will compare the new series with another one that looks more complicated but turns out to be easier to work with. In the following comparison, each term in the top series is less than its counterpart in the bottom one. Check that you believe this.

$$\frac{1}{2^2} \quad + \quad \frac{1}{3^2} \quad + \quad \frac{1}{4^2} \quad + \quad \frac{1}{5^2} \quad + \quad \frac{1}{6^2} \quad + \ldots + \quad \frac{1}{n^2} \quad + \ldots$$

$$\frac{1}{1 \times 2} + \frac{1}{2 \times 3} + \frac{1}{3 \times 4} + \frac{1}{4 \times 5} + \frac{1}{5 \times 6} + \ldots + \frac{1}{n(n+1)} + \ldots$$

This means that if the bottom series converges, then by comparison the top one must converge too. The top one isn't quite what we want because it's missing the first term $1/1^2$. But that's okay. If the top series converges to a finite number, then that finite number plus $1/1^2$ is still a finite number. So that will show that the series

$$\frac{1}{1^2} + \frac{1}{2^2} + \frac{1}{3^2} + \frac{1}{4^2} + \frac{1}{5^2} + \frac{1}{6^2} + \ldots$$

converges.

That's the outline, now the detail. The argument below uses ideas we've already explored, so if you've got this far it will be within your reach. But it is somewhat long, so you might want to read it more than once. It starts with a proof by induction about the comparator series, working with finite sums in the first instance. If you get stuck with the algebra in the middle of the induction, try working through it from bottom to top instead of top to bottom.

Claim: For every natural number n,

$$\frac{1}{1 \times 2} + \frac{1}{2 \times 3} + \ldots + \frac{1}{n(n+1)} = 1 - \frac{1}{n+1}.$$

Proof:

$\dfrac{1}{1 \times 2} = 1 - \dfrac{1}{1+1}$ so the claim is true for $n = 1$.

Suppose that the claim is true for $n = k$.

This means that $\dfrac{1}{1 \times 2} + \dfrac{1}{2 \times 3} + \ldots + \dfrac{1}{k(k+1)} = 1 - \dfrac{1}{k+1}$.

Adding $\dfrac{1}{(k+1)(k+2)}$ to both sides gives

$$\left(\frac{1}{1 \times 2} + \frac{1}{2 \times 3} + \ldots + \frac{1}{k(k+1)} \right) + \frac{1}{(k+1)(k+2)}$$

$$= \left(1 - \frac{1}{k+1} \right) + \frac{1}{(k+1)(k+2)}$$

$$= 1 - \frac{1}{k+1} + \frac{1}{(k+1)(k+2)}$$

$$= 1 - \left(\frac{1}{(k+1)} - \frac{1}{(k+1)(k+2)} \right)$$

$$= 1 - \frac{(k+2) - 1}{(k+1)(k+2)}$$

$$= 1 - \frac{k+1}{(k+1)(k+2)}$$

$$= 1 - \frac{1}{k+2}.$$

So we have established that if the claim is true for $n = k$, then it is also true for $n = k + 1$.

So, because the claim is true for $n = 1$, it must be true for every natural number n.

The induction thus establishes that for every n,

$$\frac{1}{1 \times 2} + \frac{1}{2 \times 3} + \ldots + \frac{1}{n(n+1)} = 1 - \frac{1}{n+1}.$$

And, as n gets bigger,

$$\frac{1}{n+1} \text{ gets closer to 0, so } 1 - \frac{1}{n+1} \text{ gets closer to 1.}$$

So, by a limiting argument, the infinite series

$$\frac{1}{1 \times 2} + \frac{1}{2 \times 3} + \frac{1}{3 \times 4} + \frac{1}{4 \times 5} + \frac{1}{5 \times 6} + \ldots = 1.$$

Then we can bring in the comparison.

$$\frac{1}{1^2} + \frac{1}{2^2} + \frac{1}{3^2} + \frac{1}{4^2} + \frac{1}{5^2} + \frac{1}{6^2} + \ldots$$

$$< 1 + \frac{1}{1 \times 2} + \frac{1}{2 \times 3} + \frac{1}{3 \times 4} + \frac{1}{4 \times 5} + \frac{1}{5 \times 6} + \ldots$$

$$= 1 + 1$$

$$= 2.$$

This establishes that the series

$$\frac{1}{1^2} + \frac{1}{2^2} + \frac{1}{3^2} + \frac{1}{4^2} + \frac{1}{5^2} + \frac{1}{6^2} + \ldots$$

converges to a number less than or equal to 2. That's not only finite, but really quite small. So this series might look like the harmonic series, but it behaves very differently. You might like to know that its precise sum is $\pi^2/6$.

3.11 Review

This chapter covered fraction representations, adding fractions, adding up lots of numbers, proofs without words, powers, adding up lots of numbers that happen to be expressed as powers, infinite sums, the counterintuitive property of the harmonic series, and series convergence and divergence.

It began with ways of representing and comparing fractions, considering ratios (hence *rational* numbers), equivalent fractions, and fractions as points on a number line. The number line was used to highlight differences between rational numbers and integers: rationals can be ordered, but there is no 'next number' after $\frac{1}{2}$. The early sections also highlighted situations—like those raised in Chapter 1—in which mathematical notation does not correspond in an obvious way to meaningful calculation: we can't add fractions by adding numerators and denominators, and $x^m x^n$ does not equal x^{mn} (usually). The later sections highlighted situations in which the ellipsis notation obscures important features of a series. Errors can occur if we're not alert to such things, and I suggested that it can help to check a claim against examples or to extend a series and examine more terms. The first suggestion might seem counter to the mathematical aim of constructing fully general arguments. But it isn't. General arguments might be the desired end, but mathematicians will use whatever is to hand to get there.

That said, this chapter is full of clever arguments. The sections on adding integers included reordering arguments and visual arguments— images that represent just one case but can be seen as generic. The sections on geometric sums used multiplying by the common ratio to 'shift' a sum along, so that subtracting led to lots of cancellation. We also used inductive arguments, which prove infinitely many claims by chaining them together. These highlight a different relationship between claims, examples, and generality. When we added odd integers, examples confirmed that the relevant formula seemed to work. But they gave no real insight into why—we needed a diagram or algebra for that. The inductive arguments also provided occasion to consider the relative merits of long explanations and concise proofs. If you read back now, you might find that your opinion on this has changed. And you might want to reconsider your view on genius stories and the human capacity to understand mathematics. You didn't invent the arguments in this chapter, and neither did I. But I really like them, and I hope you do too.

A final thing to learn from this chapter is that human intuitions about numbers and sums are usually based on small ones with which we have lots of experience. Sometimes these intuitions generalize to infinite sums. But sometimes they don't. This is because, as a mathematician once gleefully said to me, *infinity is really big*. It's so big that it can break our intuitions. And I argued that this is not something to worry about, but something to welcome. Broken intuitions indicate not personal failure but theoretical depth, and mathematical surprises indicate that there is a lot to learn.

CHAPTER 4

Graphs

4.1 Optimization

Here is an *optimization* problem.

A small company makes tables of two kinds, named the Hazel and the Douglas. The Hazel takes 2 days of carpentry time and 1 day of finishing time. The Douglas takes 3 days of carpentry time and 1 day of finishing time. The carpenter works for 24 days per month and the finisher works for 10 days per month. Local showrooms will buy a maximum of 8 Hazel tables and 6 Douglas tables each month. The profit on each Hazel table is £100 and the profit on each Douglas table is £120. How can the company maximize monthly profit?

I see numerous problems like this because students bring them to my university's mathematics support service. You know the phrase 'deer in the headlights'? That's what these students look like. This is clearly what teachers call a *word problem*, but it's a long way from 'If Angela has two marbles and Eric has five marbles, how many marbles do they have altogether?' Many people find the information overwhelming—they stop thinking and go into a sort of wide-eyed panic. If that happened for you, try this. Don't think about solving the problem. Just read it again, one sentence at a time, imagining the physical things described. Then look up at the ceiling and ask yourself, if you were running the company, what would you actually need to decide?

The problem requires just one decision: how many tables of each kind to make. It contains other information, but most of this is in the form of *constraints*. One constraint, for instance, is that there are 24 carpentry days per month. The company doesn't need to decide anything about that, it's just a fact. The constraints do influence the solution, though, by restricting what is possible: the company's plan must fit within the available carpentry time. The problem also contains information about profits. This will influence the solution too, though not in a straightforward way. For instance, Douglas tables make more profit. Does that mean that the company should make only those? Maybe, but maybe not. Douglas tables make more profit, but take more time. If the company made only Hazel tables, it could make more. Would that be better? How many Hazel tables could it make each month? Would the showrooms buy that many?

You might be able to reach a full solution by thinking carefully along these lines. It won't be easy, though, because the constraints are interdependent—what we really need is a way to work out how they interact. Later in this chapter we'll solve the problem by formulating a set of inequalities and relating these to a graph. After that, we'll look at other optimization problems and a variety of different graphing systems. This structure for the chapter means that if you get bogged down in the middle, it would definitely be worth skipping ahead to later sections (maybe Section 4.7 or later). But none of these ideas will make much sense unless you understand exactly what is going on in a typical graph. So we'll start with that.

4.2 Plotting points

Graphs are usually plotted on perpendicular axes, with the x-axis horizontal and the y-axis vertical. This allows any point to be specified by an x-coordinate and a y-coordinate. Often these coordinates are written as an *ordered pair* (x, y). For instance, $(2, 3)$ represents the point with x-coordinate 2 and y-coordinate 3. What are the ordered pairs for the other points marked in the following graph?[1]

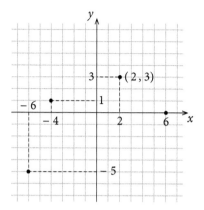

This setup permits graphing not only of individual points but also of points that are meaningfully related. For example, consider the equation $y = 2x$. Do any of the points above satisfy this equation? Which points do satisfy it?

If $x = 1$ then $y = 2 \times 1 = 2$, so $(1, 2)$ satisfies the equation;
if $x = 2$ then $y = 2 \times 2 = 4$, so $(2, 4)$ satisfies the equation, and so on.

Then there are points with zero or negative coordinates.

If $x = 0$ then $y = 2 \times 0 = 0$, so $(0, 0)$ satisfies the equation;
if $x = -1$ then $y = 2 \times (-1) = -2$, so $(-1, -2)$ satisfies the equation;
if $x = -2$ then $y = 2 \times (-2) = -4$, so $(-2, -4)$ satisfies the equation.

In fact, the points that satisfy the equation are exactly and only those on the diagonal line plotted below. At every point on this line, the y-coordinate is twice the x-coordinate. This holds whether or not the

[1] Answers: $(-4, 1)$, $(-6, -5)$ and $(6, 0)$.

coordinates are integers (recall that *integer* is the proper mathematical word for 'whole number').

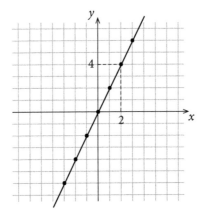

That's what it means to say that this line *is the graph of $y = 2x$*: every point on the line satisfies the equation, and every point that satisfies the equation is on the line. In my experience, people have often mastered the drawing process without really thinking about this, and we'll keep it in mind in this chapter. First, though, a word about conventions.

By convention, axes usually cross at $(0, 0)$. In the previous diagrams in this chapter, the axes extend leftward and downward, and negative numbers appear in these directions. It's common, however, to see axes drawn without the leftward and downward extensions, so that—as in the following diagrams—just an L-shape is shown. I sometimes adjust L-shaped axes when they are drawn by mathematics undergraduates, because for many mathematical situations we don't want to forget the negative numbers. But in some contexts they are reasonable. For many real-world applications, including the tables problem, only positive quantities make sense. However, the axes don't have to cross at $(0, 0)$. It can be convenient to have them frame rather than cut through a graph, which might mean that they cross at $(-1000, -1000)$, with $(0, 0)$ somewhere in the middle.

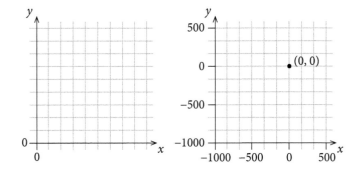

What doesn't tend to change is the arrangement of perpendicular axes with positive numbers to the right and up. Indeed, you probably learned that this is how it is—no argument, the x-axis is horizontal ('x is a-cross') with positive numbers to the right. But there is nothing mathematically essential about this, or about writing the x-coordinate first in the pair. Mathematicians could have chosen a setup with any or all of these things reversed. Or they could have gone for a libertarian system in which everyone selects an arrangement according to their mood. Operating like that, though would cost the world an enormous amount of intellectual energy. Every time anyone constructed or read a graph, they'd have to stop and think about the labelling system. People would be slower and more prone to error.

So, while exceptions are always possible, conventions are useful in mathematics as in everyday life (it's a good idea to decide that we'll all drive on the same side of the road). When accustomed to a convention we can more or less stop thinking about it, which frees up mental resources. Bear this in mind when people say that mathematics is about 'right and wrong answers'. Some things must be a certain way for mathematical reasons: once we've decided what we mean by 'regular tessellation', the only polygons that form regular tessellations are triangles, squares, and hexagons. Some things don't have to be a certain way, but for pragmatic reasons it's a good idea to make a decision and stick to it.

4.3 Plotting graphs

Here are some graphs, along with sample points, for $y = 3x$ (steeper than $y = 2x$), and for $y = x$ and $y = \frac{1}{2}x$ (both less steep).

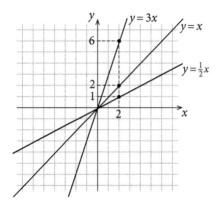

These graphs all have positive *gradients*; they slope upward from left to right. In American English one would simply say that they have positive *slopes*.[2] The gradient or slope is the number m in the equation $y = mx$, and it can be read as the ratio of vertical change to horizontal change on the graph. For $y = 3x$, that ratio is 3.

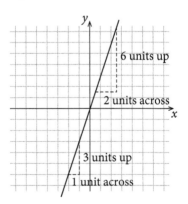

---

[2] I learned this the hard way when teaching in the USA. I used the word 'gradient' for some time before a brave student politely told me that no one in the class knew what I was talking about.

How about graphs with negative gradients? The diagram below shows graphs for $y = -3x$, $y = -x$, and $y = -\frac{1}{2}x$. If you are less familiar with graphing, check that they are drawn correctly, and compare them with their positive-gradient counterparts.

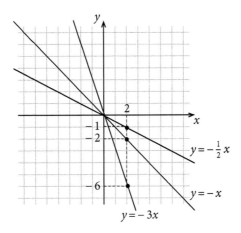

$$y = -3x$$

An equation of the form $y = mx$ is called a *linear equation*. Every linear equation has a graph that passes through $(0,0)$, because if $x = 0$ then $y = m \times 0 = 0$. It's not immediately obvious how to write equations for all possible lines through $(0,0)$, though, because so far we've seen only diagonal ones. What about the axes themselves? Can we write equations for those in the same way?

For the x-axis the answer is yes. This axis could be written as $y = 0x$, because for every value x, $y = 0x$ gives $y = 0 \times x = 0$. Alternatively, mathematicians might use the symbol '$\forall$', meaning 'for all', and write

$$y = 0 \quad \forall x.$$

In fact, they might just write

the line $y = 0$.

I find the last bit counterintuitive, because it feels like an equation describing the x-axis ought to have xs in it, not a lone y. But I can sort this out by noting that 'the line $y = 0$' refers to all points of the form $(x, 0)$, which together constitute the x-axis.

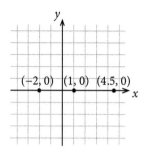

How about the vertical axis? Inspired by what we've just discussed, we could write '$x = 0 \, \forall y$', or 'the line $x = 0$'. Again I have the intuition problem, but again I can sort it out: the line $x = 0$ comprises all points of the form $(0, y)$, which together constitute the y-axis. It's worth thinking further, however, because it would be nice to write an equation for the y-axis in the form $y = mx$. Can we do that? People with good graphical intuition often want to say $y = \infty x$ (the symbol '∞' means 'infinity'). They arrive at this idea via a limiting argument. Considering $y = 2x$, $y = 3x$, $y = 4x$, and so on makes it clear that for larger values of m, equations of the form $y = mx$ have steeper graphs. Thus, it seems reasonable to say $y = \infty x$ will have 'infinite gradient', making it vertical.

This is sensible reasoning but, if you've read Chapter 3, you should hesitate. As discussed there, generalizing from the finite to the infinite sometimes works but sometimes doesn't. In this case, one problem is that a similar argument about negative gradients would lead to the conclusion that the y-axis can also be written as $y = -\infty x$. Maybe that's not a big concern if we're okay with using different expressions to specify the same line (which mathematicians are—more on that in the next section). A bigger problem is that $y = \infty x$ doesn't really give what we want. For instance, at $x = 0$, $y = \infty x$ would give $y = \infty \times 0$. What is that? You might argue that it's 0 because anything times 0 is 0. But you might argue that it's infinity because infinity times anything is infinity. Or you might argue that it's 1 because 0 and infinity in some sense 'balance out'. Does any of those arguments give what we want in terms of the graph? Unfortunately not. We'd want $y = \infty \times 0$ to somehow simultaneously give every possible number, to constitute the entire y-axis.

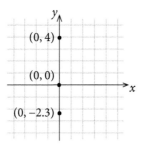

The problem is not much better elsewhere. Consider $x = 1$, for instance. If $y = \infty x$ gives anything for $x = 1$, it ought to be $y = \infty$, giving a point that is sort of 'off the top of' the graph. But this raises questions of whether $2 \times \infty$ is also ∞ and is therefore the same as $1 \times \infty$, or whether it is somehow 'twice as big'. If we allow it to be the same we get into trouble, because then we must say that

$$1 \times \infty = 2 \times \infty,$$

and dividing both sides by ∞ gives

$$1 = 2.$$

When people say that ∞ is not a number, this is what they mean—if it were a number, arithmetic would break.

If you don't know how to resolve this, that's not because your intuition is inadequate. There is no obvious answer, and mathematicians only resolved it satisfactorily towards the end of the 19th century (I'll discuss that in Chapter 5). If you don't know how to represent the y-axis in the form $y = mx$, that's because we can't. But that's okay because we can still refer to it as the y-axis or the line $x = 0$. And we can apply the reasoning about horizontal and vertical axes to other horizontal and vertical lines, which will be useful when we return to the optimization problem. For instance, the line $y = 3$ comprises all points of the form $(x, 3)$, so it is the horizontal line with y-intercept 3, as in the following left diagram. Check that you understand the labels for the vertical lines on the right.

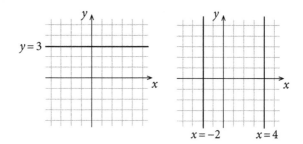

4.4 $y = mx + c$ (or b)

Those who remember graphing from school will know what's coming next: graphs for equations like $y = 2x + 3$. We can think about this equation in the familiar way.

   If $x = 0$ then $y = (2 \times 0) + 3 = 3$, so $(0, 3)$ is on the graph;
   if $x = 1$ then $y = (2 \times 1) + 3 = 5$, so $(1, 5)$ is on the graph;
   if $x = 2$ then $y = (2 \times 2) + 3 = 7$, so $(2, 7)$ is on the graph; and so on.

It's easier, though, to observe that for every value of x, the quantity $y = 2x + 3$ is three more than $y = 2x$, so the graph of $y = 2x + 3$ can be thought of as a *vertical translation* of the graph of $y = 2x$.

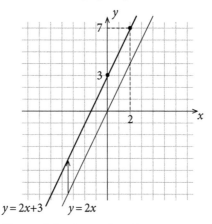

   In general, adding c shifts the graph up by c units. Because the graph then goes through the point $(0, c)$, the c is sometimes referred to as the *y-intercept*. This works with negative values of *c* too—it sounds a bit weird, but when mathematicians write a standard equation like $y = mx+c$, they usually mean that *c* could be positive or negative. This is another convention that's different in the United States, though. There, it's not $y = mx + c$, it's $y = mx + b$. I have no idea why. But this clarifies the arbitrary nature of conventions, because it clearly doesn't matter which we use. I prefer *c*, and I'd like to argue that this is because it is a good abbreviation for 'constant' (as in 'plus a constant'). But, if I'm honest, my preference is probably based on familiarity: I like the *c* because I'm used to it. In any case, here are some illustrative graphs for $y = \frac{1}{2}x$, $y = \frac{1}{2}x + 3$ and $y = \frac{1}{2}x - 1$.

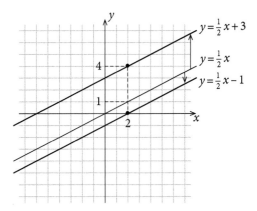

While on the subject of standard forms, I want to discuss rearranging, because an alternative will help with the optimization problem. The following algebra shows how an equation of the form $y = mx + c$ can be rearranged to put all the variables on one side. The symbol '⇔' can be read aloud as '(which) is equivalent to', and the reordering is not mathematically necessary but (unsurprisingly) it is conventional to write the *x*s first:

$$y = 2x + 3 \iff y - 2x = 3 \quad \text{(subtracting } 2x \text{ from both sides)},$$
$$\iff -2x + y = 3 \quad \text{(reordering on the left)}.$$

It's usual to end up with integer *coefficients* for each variable, which might require rearranging like this:

$$y = -\tfrac{5}{3}x + 10 \iff y + \tfrac{5}{3}x = 10 \qquad \text{(adding } \tfrac{5}{3}x \text{ to both sides),}$$

$$\iff 3y + 5x = 30 \qquad \text{(multiplying both sides by 3),}$$

$$\iff 5x + 3y = 30 \qquad \text{(reordering on the left).}$$

Hence we can rewrite $y = -\tfrac{5}{3}x + 10$ in the alternative standard form $5x + 3y = 30$. It took me a while to appreciate this because the curriculum I experienced taught me first about the form $y = mx + c$, from which you can 'read off' the gradient m and the intercept c. I got used to this and, when I learned about the alternative, I didn't find it so intuitive. But now, for graphing, I like it better. Here's why.

The equation $5x + 3y = 30$ is a rearrangement of $y = -\tfrac{5}{3}x + 10$, so it represents a straight line. And we can draw a straight line by finding two points on it and joining them together. With the $5x + 3y = 30$ form this is particularly simple. First, suppose that $x = 0$. What must y be? When x is 0, the equation $5x + 3y = 30$ becomes $3y = 30$, meaning that $y = 10$. So the point $(0, 10)$ is on the graph, and is easy to draw because it is on the y-axis. What do you think we should do next? Yep, suppose that $y = 0$. What must x be? When y is 0, the equation $5x + 3y = 30$ becomes $5x = 30$, meaning that $x = 6$. So the point $(6, 0)$ is on the graph, and on the x-axis. So we have found the two intercepts, and we can join them up.

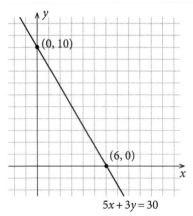

That's straightforward, and it will help with the optimization problem. But before going on I'd like to point out something fundamental. We just rearranged the equation

$$y = -\tfrac{5}{3}x + 10 \quad \text{to get} \quad 5x + 3y = 30,$$

then I confidently talked about graphing the second of these. I can do that because rearranging an equation *does not change the line to which the equation refers*. This is not immediately obvious, I think, and the first thing to do is to check that you believe it for the example discussed here. The graph should have intercept 10 and gradient $-\tfrac{5}{3}$. The intercept is easily checked, but the gradient is harder (for every one unit to the right, the graph should go five thirds of a unit down; equivalently, for every three units to the right, it should go five units down).

Why is this valid? After all, the equations $y + \tfrac{5}{3}x = 10$ and $5x + 3y = 30$ look different—the numbers in the second are all three times bigger. But suppose that the point (x, y) satisfies the first equation. Then it is true that $y + \tfrac{5}{3}x = 10$ so, multiplying both sides by 3, it is also true that $5x + 3y = 30$. This means that every point that satisfies $y + \tfrac{5}{3}x = 10$ also satisfies $5x + 3y = 30$. And the converse is true: every point that satisfies $5x + 3y = 30$ also satisfies $y + \tfrac{5}{3}x = 10$. So the two equations have exactly the same *solution set*; the set of points (x, y) that satisfies the first is exactly the same as the set that satisfies the second. And the solution set is represented by the graph, so the two equations have identical graphs.

4.5 More or less?

We've established that for an equation like $5x + 3y = 30$, the solution set is a line. Next we'll consider two related *inequalities*. First, though, a bit of mathematical pedantry. People tend to say 'equation' for any mathematical expression containing xs. But that's often inaccurate. There is an equals sign in $5x + 3y = 30$, so that is indeed an equation. But there is no equals sign in either

$$5x + 3y < 30 \quad \text{('$5x$ plus $3y$ is less than 30') or}$$
$$5x + 3y > 30 \quad \text{('$5x$ plus $3y$ is greater than 30').}$$

So these are not equations but *inequalities*. And $5x + 3y$ doesn't have a relational symbol at all—it is better described as an *expression*.

Which points satisfy the inequality $5x + 3y < 30$ and which satisfy $5x + 3y > 30$? One way to find out is to pick a point on the graph of

$5x + 3y = 30$, then think about changing x or y. The point (x, y) marked in the following is on the graph, for instance, so its coordinates must satisfy $5x + 3y = 30$. Now suppose we increase x. Then the value of $5x + 3y$ must go up, so at the new point it must be true that $5x + 3y > 30$. The same is true if we increase y.

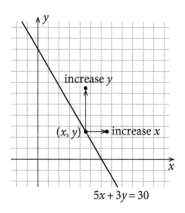

From any point on the line, moving up and/or right yields the same result. So all the points in the shaded area satisfy $5x + 3y > 30$, and all those in the unshaded area satisfy $5x + 3y < 30$.

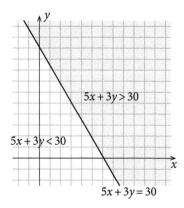

How does this work for graphs that 'lean the other way'? Consider again $y = 2x + 3$, which can be rewritten as $-2x + y = 3$ (check that you believe this). How does the sketching process give the following graph, and can you convince yourself that the inequalities are on the correct sides of the line? What difference does the negative x-coefficient make?

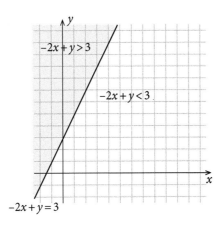

With this reasoning about inequalities in place, we're ready to revisit the opening optimization problem. Here it is again.

A small company makes tables of two kinds, named the Hazel and the Douglas. The Hazel takes 2 days of carpentry time and 1 day of finishing time. The Douglas takes 3 days of carpentry time and 1 day of finishing time. The carpenter works for 24 days per month and the finisher works for 10 days per month. Local showrooms will buy a maximum of 8 Hazel tables and 6 Douglas tables each month. The profit on each Hazel table is £100 and the profit on each Douglas table is £120. How can the company maximize monthly profit?

Recall that the company needs to decide how many tables of each kind to make. It's often a good idea to assign variable names to things we want to investigate, so let's say that the company makes h Hazel tables and d Douglas tables. As noted earlier, one constraint is the amount of carpentry time, capped at 24 days. Each Hazel table takes two carpentry days, and each Douglas table takes three. So making h Hazel tables and d Douglas tables takes $2h + 3d$ days in total, and the company must ensure that

$$2h + 3d \le 24.$$

This type of inequality is familiar. The letters are not x and y, but that's not a big problem. We could change h and d to x and y, but x and y don't remind me of the table types, so if we did that I'd likely get mixed up. Instead, I'd prefer to relabel the axes in a standard graph. If we do that, we can represent this inequality as below. The shaded region contains all points ruled out by the constraint, the white region and the boundary show solutions that are *feasible*.

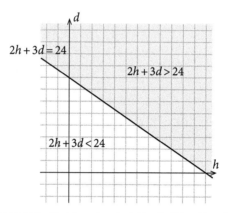

The next constraint, for finishing time, has inequality is $h + d \leq 10$ (why?). The two constraints together are shown in the following, and the *feasible region* is now a bit smaller.

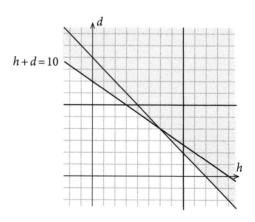

Finally, the showrooms will buy only 8 Hazel tables and 6 Douglas tables, so the company must ensure that

$$h \leq 8 \quad \text{and} \quad d \leq 6.$$

The next graph adds these, together with two final, real-world constraints: the company will not be making negative numbers of tables, so both h and d must be greater than or equal to 0. The feasible region is now quite small, and the optimal solution—the one that maximizes profit—must be inside or on the boundary of the white shape. Before you read on, where you do think it is? If you can't be specific, can you say roughly where it is likely to be and why?

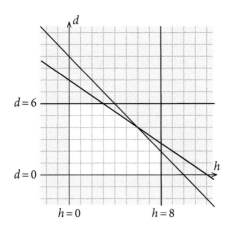

4.6 Intersecting lines

More tables means more profit, so the optimal solution will be where h and d are large: at the top of the feasible region (where d is maximal), at the right (where h is maximal), or in some top-right-ish position that balances both. For more precision, we will think about the labelled points in the following graph. This is the final section on this problem, so remember what I said at the beginning of the chapter: if you get bogged down here, skip to the next section for a restart on different ideas.

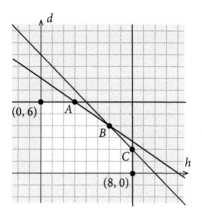

To find the optimal solution, we need to know how to calculate profit. Because the profit on each Hazel table is £100 and the profit on each Douglas table is £120, making h Hazel tables and d Douglas tables yields

$$\text{profit (in £)} = 100h + 120d.$$

In the previous graph, the point $(0, 6)$ corresponds to making no Hazel tables and 6 Douglas tables, which yields

$$\text{profit (in £) at } (0, 6) = 100 \times 0 + 120 \times 6 = 720.$$

Can that be optimal? No, because making 6 Douglas tables leaves spare carpentry, finishing, and showroom capacity. To find out how many Hazel tables the company could make alongside the 6 Douglases, we'd like to know the position of point A. Probably you can just read this because I've made the graph accurate. But we can check algebraically. The point A is where the line $d = 6$ intersects the line $2h + 3d = 24$. That means that it satisfies *both equations at once*. So $d = 6$, and substituting this into $2h + 3d = 24$ gives $2h + 18 = 24$, meaning that $2h = 6$, so $h = 3$. So the point A is $(3, 6)$, which corresponds to making three Hazel tables and six Douglas tables and gives

$$\text{profit (in £) at } (3, 6) = 100 \times 3 + 120 \times 6 = 1020.$$

Similar reasoning shows that point C has coordinates $(8, 2)$, with

$$\text{profit (in £) at } (8, 2) = 100 \times 8 + 120 \times 2 = 1040.$$

That's slightly better. So, of these options, the company should choose to make 8 Hazel and 2 Douglas tables.

How about the remaining top-right-ish points? The point B is at the intersection of the lines $2h + 3d = 24$ and $h + d = 10$. So its coordinates satisfy both of:

$$2h + 3d = 24 \quad (1)$$
$$h + d = 10 \quad (2)$$

Can you solve these *simultaneous equations*? Can you do it in more than one way? A serviceable but inelegant approach is to rearrange equation (2) to give $h = 10 - d$, then replace h with $10 - d$ in equation (1). Try this if you like. A more elegant approach is the one that you probably learned in school, for which I'd write something like this. Can you explain why each step is both valid and sensible?

Let	$2h + 3d = 24$	(1)
and	$h + d = 10$	(2)
Multiplying (2) by 2 gives	$2h + 2d = 20$	(3)
Subtracting (3) from (1) gives	$d = 4$	
Substituting $d = 4$ into (2) gives	$h = 6.$	

This works because multiplying an equation by a nonzero number doesn't change its solution set, so equations (2) and (3) have the same solution sets. Subtracting (3) from (1) subtracts 20 from both sides (we don't know the values of h and d at this stage, but we know that two of each adds up to 20). This yields the coordinates $(h, d) = (6, 4)$, and

profit (in £) at $(6, 4) = 100 \times 6 + 120 \times 4 = 1080.$

So option B is now the profit winner. But what about the other boundary points, those on an edge of the feasible region but not at a corner? Some of these are not practically possible: the company will not be making four and half tables. But might a possible edge point be better? Or a point just inside the region, perhaps? Maybe try out some calculations.

In fact, point B is the best solution. In problems of this type, there will always be an optimal solution at a corner point of the feasible region (any corner, not necessarily a right-angled one). We can think about why like this. We established earlier that at the point $(0, 6)$, the profit is £720. Which other points give the same profit? To answer this, we'd like to find out where

$$100h + 120d = 720.$$

That's a familiar type of equation and its graph will be a straight line. We already know one point on this line, $(0, 6)$, so we can draw the whole line by finding one more. On the x-axis, for instance, $100x = 720$, so $(7.2, 0)$ is on the graph, as shown in the following graph. Every point on the dotted line has the same profit, £720.

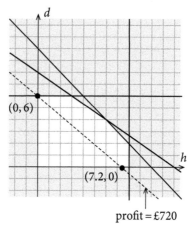

profit $= £720$

Similarly, point A is $(3, 6)$, and we established earlier that the profit at A is £1020. Other points giving the same profit will satisfy $100h + 120d = 1020$, and this graph will have the same gradient as the one we just drew. Why, exactly? Both graphs appear next.

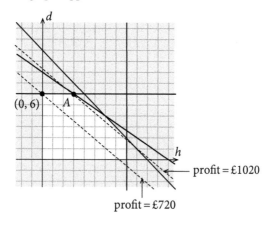

profit $= £1020$

profit $= £720$

Now, what would the graphs look like for other profit lines? Can you see why *B* must be the best solution? And can you imagine having more than one optimal solution? To help with that, here is another problem. If you fancy it, have a go at the whole solving process for yourself.

A small company makes two kinds of decorated cake, named the Cameron and the Iona. The Cameron takes 1 hour of baking time and 1 hour of icing time. The Iona takes 1 hour of baking time and 2 hours of icing time. The baker works for 6 hours per day and the person who does the decorative icing works for 8 hours per day. A local market stall will buy a maximum of 5 Cameron cakes and 3 Iona cakes per day. The profit on each cake of either type is £5. How can the company maximize daily profit?

4.7 Areas and perimeters

The next part of this chapter will look at different types of optimization problem. We'll start with a classic.

A farmer has 20 metres of fencing, and he plans to build a rectangular enclosure. He wants the area of the enclosure to be as large as possible. How should he do it?

In this problem 20 metres is the *perimeter* of the enclosure—the combined length of its edges. One appropriate rectangle would have two 4-metre edges and two 6-metre edges; its area would be $4 \times 6 = 24$ square metres.

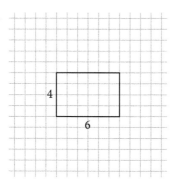

Cynics will note that this is one of those problems where the context is just window dressing. What it really says is

Of all the rectangles with perimeter 20, which has the largest area?

Does the 4 × 6 rectangle have the largest area? There's thinking to do here, because for some people it's not obvious that the question makes sense. One natural intuition is that shapes with the same perimeter are the same 'size', so they have the same area. But 'size' does not have an obvious interpretation in two dimensions; perimeter and area do not change in the same way. For instance, doubling the edges of the 4 × 6 rectangle doubles its perimeter but quadruples its area.

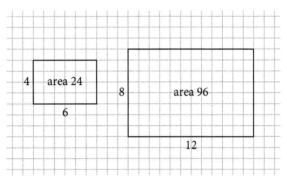

More importantly for the farmer's problem, fixing a rectangle's perimeter does not fix its area. The rectangles below all have perimeter 20, but their areas differ—the biggest area is more than twice that of the smallest.

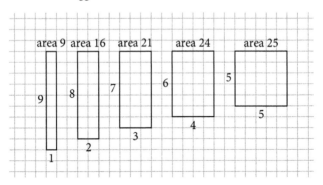

Based on these rectangles it appears that the farmer should build a square enclosure. But how sure would you be? Could there be a better alternative with noninteger edge lengths? If you think that the square is optimal, is that because the perimeter is 20? Or would a square be best for every perimeter? You might want to play around to get a feel for this—maybe try a smaller perimeter, like 12, and a bigger one, like 100.

In fact, the square *is* always the best option, and this can be understood at various levels of sophistication. One is an argument based on symmetry. Fixing the perimeter means that as the width of the rectangle increases, its height decreases. Tabulating some values shows a growing-then-shrinking pattern in the areas.

Width	Height	Area
1	9	9
2	8	16
3	7	21
4	6	24
5	5	25
6	4	24
7	3	21
8	2	16
9	1	9

Tabulated values don't explain why the biggest area occurs when the width and height are equal, though. To think about that, visual representations help. Suppose we start with the 5 × 5 square. Shrinking the height by one and extending the width by one keeps the perimeter constant. How does it affect the area?

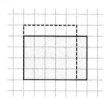

This change slices off a row of five square units and replaces them with a column of four square units, meaning that the area shrinks by one square unit. What happens for other widths and heights in the table? What would happen for a smaller perimeter like 12 or a larger one like 100? What if we start with the 5 × 5 square and change the height and width by $\frac{1}{2}$, or by $\frac{1}{10}$? Does this explain why the square is always optimal?

Mathematically, I think the fact that the square is optimal is a wonderfully elegant result. A square is a special rectangle not only because it has extra symmetry; it is also, for a given perimeter, the rectangle with largest area. With that in mind, consider this related problem.

> A farmer has 20 metres of fencing, and he plans to build a rectangular enclosure *where one side of the enclosure is formed by his farmhouse wall.* He wants the area of the enclosure to be as large as possible. How should he do it?

What do you think? This time the farmer needs fencing for only three sides (we're supposed to assume that his wall is long enough to form one entire side of the rectangle). Does that make a difference? Will the optimal solution still be a square? Before reading on, you might want to experiment with some length options, find the areas, and see whether your intuition seems to be correct.

4.8 Area formulas and graphs

The previous section might have seemed to wander away from the nominal topic of this chapter: graphs. But graphs remain useful and, although the shapes in these problems are rectangular, the relevant graphs are curved. This is because the enclosure edge lengths are not independent. In the first problem, for instance, if one edge has length 1, its neighbour must have length 9. That's because, if the perimeter is 20, any pair of adjacent edges must add up to 10. Algebraically, if one edge has length x, the adjacent one must have length $10 - x$, meaning that the area is $x(10 - x)$.

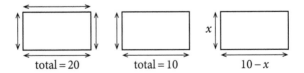

For the second problem, if an edge perpendicular to the wall has length x, then the edge parallel to the wall must have length $20 - 2x$ (the remaining fencing is used for one edge, not two). So the area is $x(20 - 2x)$. Now, do you want to reconsider your answer about the optimal enclosure? Is it still a square?

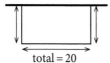 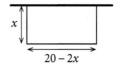

Tabulating values again highlights relevant patterns.

x	$10 - x$	$x(10 - x)$		x	$20 - 2x$	$x(20 - 2x)$
1	9	9		1	18	18
2	8	16		2	16	32
3	7	21		3	14	42
4	6	24		4	12	48
5	5	25		5	10	50
6	4	24		6	8	48
7	3	21		7	6	42
8	2	16		8	4	32
9	1	9		9	2	18

All the area entries in the second table are bigger. This isn't surprising: using the fencing for three sides rather than four should enclose more area. In fact, the entries in the second table are double their counterparts in the original. We can see why by rearranging the second formula.

$$\text{area (4 sides)} = x(10 - x);$$
$$\text{area (3 sides)} = x(20 - 2x) = 2x(10 - x).$$

Finally, is the optimal solution still a square? No, it's not—the optimal solution still has $x = 5$, but the corresponding shape is now a rectangle in which the edge parallel to the wall is double that length. You might want to develop your intuition for this by thinking as we did before about increasing and decreasing lengths and widths. More general insight is then available by plotting graphs to show how area varies with x. The areas, represented as y-values in the following graphs, are considerably bigger than the x-values, so I've adjusted the axis scales. One curved graph shows $y = x(10 - x)$ and the other shows $y = x(20 - 2x)$. Which is which?

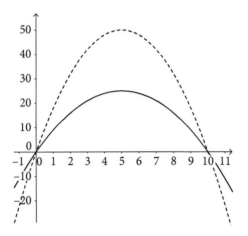

These graphs go beyond both our reasoning and the original problem. They go beyond our reasoning by showing the y-values for noninteger x-values and for values beyond those tabulated. At $x = 0$ and $x = 10$, for instance, the y-value is 0, so the area would be 0. Why is that, in relation to the problem? And the graphs allow $x < 0$ and $x > 10$. The problem makes no sense if x is negative or if $(10 - x)$ is negative—there is no such thing as a negative-length fence. But the equation $y = x(10 - x)$ requires no such restriction. For instance, $x = -1$ gives $y = (-1)(10 - (-1)) = (-1).11 = -11$. Where is this on the graph?

People who have studied *quadratic functions* will recognize these graphs as *parabolas*. We haven't written their equations in the standard form, but we could. For instance,

$$y = x(20 - 2x) = 20x - 2x^2 = -2x^2 + 20x.$$

This is an instance of the standard form of a quadratic equation

$$y = ax^2 + bx + c,$$

with $a = -2$, $b = 20$, and $c = 0$. Why do you think quadratic equations are so called?

To conclude this section, how about inverting the problems? Suppose that instead of maximizing area for a fixed amount of fencing, the farmer wants to minimize the amount of fencing to enclose a fixed area. Then the two problems might read like this.

A farmer wants to build a rectangular enclosure with area 100 square metres. He wants to use the minimal amount of fencing. How should he do it?

A farmer wants to build a rectangular enclosure with area 100 square metres where one side of the enclosure is formed by his farm-house wall. He wants to use the minimal amount of fencing. How should he do it?

The information so far will allow you to guess the correct enclosure 'shapes'. But you might like to work through some tables, formulas, and graphs to join everything up.

4.9 Circles

Quadratic equations give graphs with one type of curve: a parabola. Another obvious curve to consider is a circle. Have you ever wondered about how to write an equation for a circle? It is not possible do this in the form 'y = some function of x', in part because a circle curves back on itself: points such as x_1 in the following graph have no corresponding y-values. That's not mathematically serious because we could restrict the *domain*, defining the relevant function only for certain x-values. More seriously, points such as x_2 have two distinct corresponding y-values. So we can't write y as a function of x, because at some points we'd want y to somehow be two different numbers.

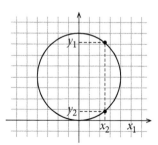

That probably worries mathematicians more than it worries the typical reader of this book—if you ruled the world you might be inclined to allow *multivalued* functions. In fact, though, it is possible to write an equation for a circle, just in a different form. It's easiest to centre the circle at the *origin* $(0, 0)$; this makes everything symmetrical. Then, pleasingly, an equation can be constructed using Pythagoras' theorem, as discussed in Chapter 1. This might be surprising, because Pythagoras' theorem is about triangles, not circles. But judicious use of triangles helps a lot.

The key is that a circle, whatever its size, has a fixed *radius*: the distance from the centre to a point on the circle is always the same. Say the radius is r, and take a point (x, y) on the circle. Drop a perpendicular from (x, y) to the x-axis, and note that doing this constructs a right-angled triangle with one edge of length x, one of length y, and hypotenuse r. What does Pythagoras' theorem say about this triangle? It says that $x^2 + y^2 = r^2$. This will be true for every point on the circle, so the equation of the circle is $x^2 + y^2 = r^2$.

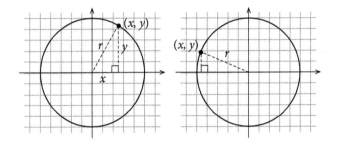

If you've a mathematical turn of mind, this might worry you a bit. Not because there's anything wrong with it—there isn't—but because the x and y values can be negative or 0. A mathematically alert person ought to worry about whether everything still works in those cases, so we'll check. At the point $(0, r)$ in the following graph, for instance,

$$x = 0 \text{ and } y = r, \quad \text{so } x^2 = 0 \text{ and } y^2 = r^2, \quad \text{so } x^2 + y^2 = r^2.$$

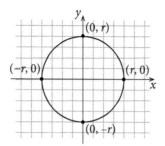

How about the point $(0, -r)$? That's okay too. This time

$x = 0$ and $y = -r$,   so $x^2 = 0$ and $y^2 = (-r)^2 = r^2$,   so $x^2 + y^2 = r^2$.

If the fact that $(-r)^2 = r^2$ bothers you because you've never really understood why 'a minus times a minus is a plus', trust me for now and you'll find an explanation in Chapter 5. But it's worth another word here about reading mathematics aloud. Experienced mathematicians do this flexibly, and sometimes use their knowledge to read beyond the literal. For instance, it would be perfectly accurate to read

$x^2 + y^2 = r^2$ as   'x squared plus y squared equals r squared'.

But a mathematician might well read

$x^2 + y^2 = r^2$ as   'the circle of radius r centred at the origin'.

Expertise is an interesting thing.

Having established $x^2 + y^2 = r^2$ as a general equation for a circle, it's educative to vary it. First, we could have bigger or smaller r. Here are the graphs for $x^2 + y^2 = 4$ and $x^2 + y^2 = 16$.

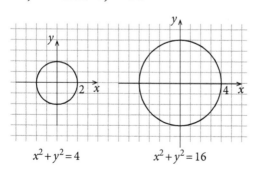

$x^2 + y^2 = 4$ $x^2 + y^2 = 16$

Second, what shape would be given by this variant?

$$4x^2 + y^2 = 16.$$

I'll give you a clue: it's one of the following *ellipses*. Can you work out which?

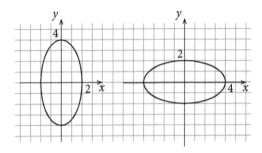

It's the one on the left, which I find counterintuitive because the equation $4x^2 + y^2 = 16$ makes me think that the ellipse should be bigger in the x direction. In fact, though, the coefficient of x^2 being 4 has the opposite effect. I haven't overcome the faulty intuition—for me this is like 'multiplication makes things bigger' all over again. But I have developed the sense to stop each time and think about axis crossings. Where $x = 0$, the equation $4x^2 + y^2 = 16$ becomes $y^2 = 16$, so $y = \pm 4$ (the symbol '$\pm$' is read aloud as 'plus or minus'). Where $y = 0$, however, $4x^2 + y^2 = 16$ becomes $4x^2 = 16$, meaning that $x^2 = 4$, so $x = \pm 2$. So the crossings on the x-axis are closer to 0, and the circle is squashed into a tall ellipse.

Finally, how about a circle with a different centre? In the following diagram, the radius is still r but the centre of the circle is at the point $(5, 2)$. This means that the horizontal edge of the labelled triangle has length $x - 5$, and the vertical edge has length $y - 2$.

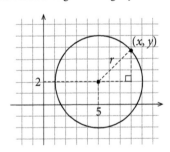

Applying Pythagoras' theorem as before, this means that every point (x, y) on the circle satisfies the equation

$$(x - 5)^2 + (y - 2)^2 = r^2.$$

And a mathematician might well read this equation as 'the circle of radius r centred at the point $(5, 2)$'. If you find it hard to imagine ever developing the knowledge to just recognize that, think about your own mathematical learning over time. Probably you recognize $y = mx + c$ as a straight line, maybe even as a straight line with gradient m and y-intercept c. But the 10-year-old you would have found that amazing.

4.10 Polar coordinates

I'll conclude the main content of this chapter with two sections on representation systems. The *Cartesian* coordinate system used so far is now ubiquitous. It is named for the 17th-century mathematician Descartes, though others used and adapted similar systems. This might give some readers another historical shock. The familiar coordinate system was indeed invented. Someone, at some point, had to say 'Hey, you know how we keep wanting to represent the way in which values of one quantity are related to values of another? Why don't we put the quantities on perpendicular axes so that each pair of values gives a single point?' It's hard to imagine this now because Cartesian coordinates are part of the mathematical furniture—they're so familiar that we don't really think about them.

It happens, though, that Cartesian coordinates are not the only way of representing points in two dimensions. Two dimensions require two coordinates, but these do not have to be x and y coordinates based on horizontal and vertical axes. One alternative is *polar coordinates*, where *polar* refers to a *pole* or central point from which we measure, as in 'north pole' (though not on a sphere). Polar coordinates are usually notated (r, θ), where r is distance from the origin, and θ (the Greek letter *theta*) is the angle measured anticlockwise from what would be the x-axis.

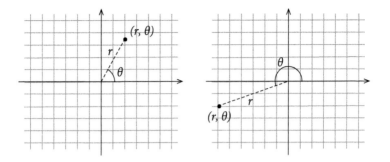

This system has a key feature in common with Cartesian coordinates: it allows us to specify every point. But it differs in another key feature: the names of points are no longer unique. For instance, the following point could be specified in either of the two given ways.

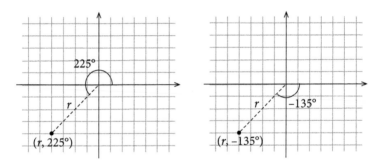

This can be bothersome in practice, so people tend to adopt a convention in which all stated angles are between either 0° and 360° or –180° and 180°. But it's not considered bothersome in principle, because multiple expressions for the same object are mathematically common. As we've seen, the fractions $\frac{1}{2}$ and $\frac{3}{6}$ both specify the same point on the number line, and the two equations $y + \frac{5}{3}x = 10$ and $5x + 3y = 30$ both specify the same line. Here, the two ordered pairs $(r, 225°)$ and $(r, -135°)$ specify the same point.

Polar coordinates are useful because Cartesian coordinates are untidy for circles: an equation like $x^2 + y^2 = r^2$ isn't especially elegant. In polar coordinates the equation is much nicer. A circle is all the points at distance

r from the origin or pole, so the polar equation for a circle is simply $r = 2$ or $r = 4$ or similar.

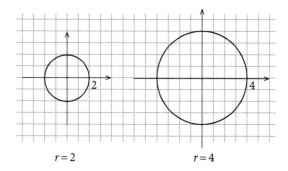

$r = 2$ $r = 4$

This equation does not involve θ because for every value of θ, the value of r is the same. We could write '$r = 2 \ \forall \theta$', but we don't have to. In polar coordinates, it's fine to write $r = 2$ to represent a circle, just as in Cartesian coordinates it's fine to write $x = 3$ to represent a vertical line. And, as in Cartesian coordinates, we can specify regions using inequalities. For instance, the region $r \leq 2$ is shaded in the following left diagram. What does the region $r \geq 2$ look like? And the region $r > 2$? We can specify an *annulus* with, for instance, one boundary included and one not, by the double inequality $2 \leq r < 4$ (read aloud as '2 is less than or equal to r is less than 4' and meaning that both $2 \leq r$ and $r < 4$).

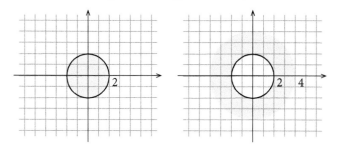

We can specify regions in terms of θ, too. For instance, the following left region is specified by $0 \leq \theta \leq 60°$. And the right region is specified by the two inequalities $0 \leq \theta \leq 60°$ and $2 \leq r \leq 4$. It makes sense to think of the latter as a sort of 'polar rectangle'. Can you see why?

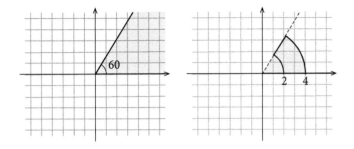

4.11 Coordinates in three dimensions

To conclude this chapter, we'll consider coordinate systems in three dimensions. To extend Cartesian coordinates, we commonly imagine a third axis perpendicular to both the x- and y-axes, and labelled z. Equations then slice up three-dimensional space. But now, instead of a line, an equation like $x = 3$ represents a *plane*. That's because $x = 3$ refers to all points of the form $(3, y, z)$, where y and z can be anything you like.

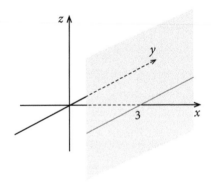

For a more complicated example, consider the equation $x + y + z = 2$. We can work out what this looks like by thinking about axis crossings (handy approach, that). When y and z are both 0, x must be 2. So the point $(2, 0, 0)$ is on the plane, as are $(0, 2, 0)$ and $(0, 0, 2)$. So this plane passes through all of these points.

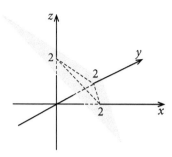

And three-dimensional solids can be specified by combining inequalities. The region that satisfies all three of $0 \leq x \leq 3$, $0 \leq y \leq 2$, and $0 \leq z \leq 1$ is a *cuboid* or *rectangular prism*. In the left diagram below, every point in the box has x-coordinate between 0 and 3, y-coordinate between 0 and 2 (the 2 isn't labelled because it's 'at the back of' the box), and z-coordinate between 0 and 1. The right diagram shows the solid that satisfies all four of $x \geq 0$, $y \geq 0$, $z \geq 0$, and $x + y + z \leq 2$. This is a *tetrahedron* or *triangular pyramid*—can you work out how it corresponds to the inequalities?

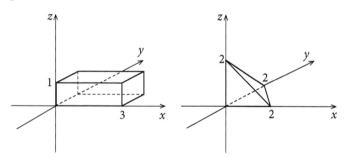

Now, how do you think we might extend polar coordinates? This is not obvious and there are two commonly used systems, termed *cylindrical polar coordinates* and *spherical polar coordinates*. Cylindrical polars just insert a z-axis perpendicular to the polar plane, so that (r, θ, z) is r units from the origin, an angle of θ anticlockwise from the x-axis, and z units up. This system, unsurprisingly, is good for representing cylinders. The equation $r = 2$, for instance, specifies an infinite cylinder of radius 2.

The inequalities $r \leq 2$ and $0 \leq z \leq 1$ together specify a solid cylinder of height 1.

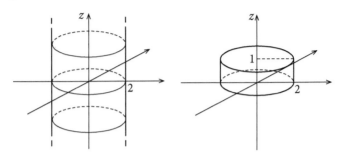

Spherical polars are a bit harder to think about, at least for me. A spherical polar point is usually denoted (r, θ, ϕ), where ϕ is the Greek letter *phi*. This point is r units from the origin, an angle θ anticlockwise from the x-axis, and an angle ϕ down from the vertical. If you don't mind looking like a Dalek,[3] you can generate physical intuition for this by thinking of one of your shoulders as the origin $(0, 0, 0)$. Stick your arm straight up and pretend its length is r. Then keep it straight and sweep it downwards—by doing that you're increasing ϕ. Then keep your shoulder still and keep your hand at the same height but swing it around—by doing that you're changing θ.

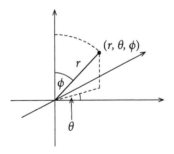

Spherical polars are, of course, good for representing spheres. In spherical polars, the equation $r = 3$ specifies a hollow sphere of radius 3.

---

[3] Daleks are cyborg aliens from the UK science-fiction TV show *Doctor Who*.

If you're not sure why, fix your shoulder in place, swing your straight arm around and think about all the points your fingertips would reach. The inequalities $r \leq 3$ and $\phi \leq 45°$ together specify a sort of fat solid ice-cream cone.

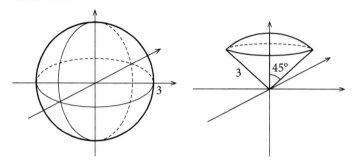

To finish off, you might like to test your understanding of both co-ordinate systems by working out what three-dimensional shapes would be specified by these sets of inequalities:

(cylindrical) $2 \leq r \leq 4$ and $-2 \leq z \leq 2$;
(cylindrical) $r \leq 4$ and $\theta \leq 90°$ and $0 \leq z \leq 2$;
(spherical) $r \leq 2$ and $\theta \leq 90°$;
(spherical) $r \leq 2$ and $\theta \leq 180°$ and $\phi \leq 90°$.

And you might like to note that different coordinate systems are not just an exercise in pure mathematics, they have enormous practical utility. We live on a sphere and, while we don't notice this at the everyday human scale, we have technologies for which it is very relevant. To fly jets between continents or track satellites in orbit, spherical polar coordinates are perfect.

4.12 Review

This chapter discussed optimization, graphs, and links between the two. It began with plotting individual points, then progressed to graphing linear equations. We considered representing axes—highlighting the fact that ∞ is not a number—and writing equations in different standard forms. We then related graphs to inequalities. In the later sections, we considered classic areas-and-perimeters problems, relating the optimal solutions to curved graphs of quadratic functions. Finally, we looked at different coordinate systems for both two-dimensional and three-dimensional space, and at regions specified by inequalities.

One implicit theme was using graphs to facilitate problem solving by representing lots of information at once. An equation like $y = 2x + 3$ has an infinite solution set, and the corresponding graph captures this set in a convenient form. It also divides the plane into two regions where, respectively, $y > 2x + 3$ and $y < 2x + 3$. Interpreting such regions allowed us to solve the opening optimization problem by representing all the relevant information on a single graph.

A more explicit theme was conventions, which standardize communication and minimize collective effort. I contrasted arbitrary (but useful) conventions with necessary mathematical consequences. For instance, we can choose to represent points using either Cartesian or polar coordinates. But once we've decided, a circle centred at the origin must have equation $x^2 + y^2 = r^2$ in Cartesian coordinates, and $r = a$ in polars. The communication benefit is immediate because standard equation types have standard graphs. Sketching might take a while, but in Cartesian coordinates I know that $y = 3x - 29$ will be a straight line, $y = x^2 - 3x$ will be a parabola, and $x^2 + y^2 = 196$ will be a circle. And this facilitates spoken communication. Saying 'the circle of radius 3 centred at the origin' is no more accurate than saying 'x squared plus y squared equals 9', but for listeners it probably creates a more vivid mental image. And imagery can help in all kinds of problem solving. People with well-integrated knowledge of different representation systems can translate flexibly between them, using their respective advantages. Fluent translation is a valuable mathematical skill.

CHAPTER 5

Dividing

5.1 Number systems

What is a number? This question is far from trivial—numbers are slippery beasts. For instance, the numeral '5' can be thought of as a number, but it's not really. It's fair to say that it's a canonical representation of the number five, but five can also be written in numerous other ways.

$$5 \qquad 1 \times 5 \qquad 5.00 \qquad \text{five} \qquad \frac{30}{6}$$

$$4 + 1 \qquad \therefore \qquad \sqrt[3]{125} \qquad \text{卌}$$

$$\sqrt{25} \qquad 8 - 3 \qquad 10 \div 2 \qquad (-1) \times (-5)$$

These expressions all represent the same abstract object—if they refer to a thing, then five is that thing. But they represent it in different ways, in some cases highlighting links to broader number systems. In this chapter, those systems are a major theme.

To see why that's important, consider another question. Why do numbers work the way they do? In a limited sense, this is due to regularities in the world. If you have three sandwiches and I have two sandwiches, then altogether we have five sandwiches. It's worth having names for 'two', 'three', and 'five' because the same pattern would hold if instead we had toothbrushes or candlesticks. And other mathematical concepts are grounded in other perceptual experiences. You know what a straight line is, and what symmetry is, and you probably feel that these notions are natural features of the world, rather than just in your head. But mathematics climbs fast towards abstraction. Why, for instance, was it okay in Section 3.6 for me to explain that 2^0 can be defined by generalizing a relationship that makes sense for positive powers? There's nothing 'in the world' that directly informs this. It's less about numbers as separate things and more about number systems.

This final main chapter will discuss logical consequences of mathematical relationships and how these are represented in number systems. With that in mind, here is something to start us off. Did you know that a number is divisible by 9 if and only if the sum of its digits is divisible by 9? For example:

The sum of the digits of 738 is $7 + 3 + 8 = 18$, which is divisible by 9. So 738 is divisible by 9 (check: $738 \div 9 = 82$).

The sum of the digits of 61 237 is $6 + 1 + 2 + 3 + 7 = 19$, which is not divisible by 9. So 61 237 is not divisible by 9 (check: $61\,237 \div 9 = 6804.111\ldots$ or, if you prefer, $61\,237 \div 9 = 6804$ remainder 1).

This is cute and it makes for a clever-looking party trick. But the interesting question is, why does it work? Such regularities don't come out of nowhere. This one arises because we represent numbers in *base* 10. We'll start by exploring base-10 representation in relation to this phenomenon.

5.2 Dividing by 9 in base 10

To say that we represent numbers in base 10 means that the places in our *place-value* number system correspond to powers of 10. You might have learned to describe this in terms of hundreds, tens, and units. For instance, the number 738 has 7 in the hundreds place (where $100 = 10^2$), 3 in the tens place ($10 = 10^1$), and 8 in the units place ($1 = 10^0$; if you're not sure why, see Section 3.6):

hundreds tens units

7   3   8

This makes it possible to consider what changes from one multiple of 9 to the next. Adding 9 is the same as adding 10 and subtracting 1. So, at each step in the list below, the tens digit goes up by 1 and the units digit goes down by 1. This keeps the digit sum constant so, because the first number is 9, the total at every step is 9.

$$9$$
$$18$$
add 1 ten $\Big\langle \begin{matrix} 27 \\ 36 \end{matrix} \Big\rangle$ subtract 1 unit
$$45$$
$$54$$
$$63$$
$$72$$
$$81$$
$$90$$

Unfortunately, this works only until 90, after which it gets more complicated. Adding 9 to 90 gives 99, leaving the tens digit unchanged and adding 9 to the units digit. So the total remains divisible by 9. Adding 9 to 99 gives 108, which drops the tens digit by 9, drops the units digit by 1, and increases the hundreds digit by 1—again the total remains divisible by 9. Then it's back to adding a 10 and subtracting a unit.

$$90 \;\rangle$$ add 9 units
add 1 hundred and subtract 9 tens $\Big\langle \begin{matrix} 99 \\ 108 \end{matrix} \Big\rangle$ subtract 1 unit
add 1 ten $\Big\langle \begin{matrix} 108 \\ 117 \end{matrix} \Big\rangle$ subtract 1 unit

Similar reasoning might convince you of the general result that a number is divisible by 9 if and only if the sum of its digits is divisible by 9 (what happens at the next hundred or thousand?). But greater insight is possible if we think less about the adding process and more about the number's structure. Consider again 738, which we could write as

$$738 = (7 \times 100) + (3 \times 10) + (8 \times 1).$$

This helps because although neither 100 nor 10 is divisible by 9, both are very close to something that is: we can replace 100 with 99 + 1 and 10 with 9 + 1.

$$738 = (7 \times 100) \quad + (3 \times 10) \quad + (8 \times 1)$$
$$= (7 \times (99 + 1)) + (3 \times (9 + 1)) + (8 \times 1).$$

Then we can multiply out using distributivity, reorder the terms using commutativity, collect all the '×1' bits together, and write them more simply. Make sure you can see what happens in each new equation in this array.

$$738 = (7 \times 100) \quad\quad + (3 \times 10) \quad\quad + (8 \times 1)$$
$$= (7 \times (99 + 1)) \quad + (3 \times (9 + 1)) \quad + (8 \times 1)$$
$$= (7 \times 99) + (7 \times 1) + (3 \times 9) + (3 \times 1) + (8 \times 1)$$
$$= (7 \times 99) + (3 \times 9) + (7 \times 1) + (3 \times 1) + (8 \times 1)$$
$$= (7 \times 99) + (3 \times 9) + ((7 + 3 + 8) \times 1)$$
$$= (7 \times 99) + (3 \times 9) + (7 + 3 + 8).$$

Finally, think about the three parts of the last expression.

$$738 = \underbrace{(7 \times 99)}_{\text{divisible by 9}} + \underbrace{(3 \times 9)}_{\text{divisible by 9}} + \underbrace{(7 + 3 + 8)}_{\text{divisible by 9}}.$$

This shows that 738 is divisible by 9 because it is equal to a sum in which all addends are divisible by 9. The first and second addends are multiples of 99 and 9, respectively. The third is a multiple of 9 too, which is easy to check by adding. But what is the third addend? It's the sum of the digits.

Moreover, for different three-digit numbers, the third addend will always be the sum of the digits. For instance, 655 is not divisible by 9.

$$655 = \underbrace{(6 \times 99)}_{\text{divisible by 9}} + \underbrace{(5 \times 9)}_{\text{divisible by 9}} + \underbrace{(6 + 5 + 5)}_{\underline{\text{not}} \text{ divisible by 9}}.$$

Are you convinced, or would it help to write out the full argument for this number? If you can see these arguments as generic, that's good. But mathematicians especially like fully general arguments, and here we can use a letter to stand for each digit. Suppose that the number is $d_2d_1d_0$, where the letters are not numbers multiplied together, but digits, meaning

$$
\begin{array}{ccc}
\text{hundreds} & \text{tens} & \text{units} \\
d_2 & d_1 & d_0
\end{array}
$$

Then the argument has exactly the same structure.

$$
\begin{aligned}
d_2d_1d_0 &= (d_2 \times 100) & + (d_1 \times 10) & + (d_0 \times 1) \\
&= (d_2 \times (99 + 1)) & + (d_1 \times (9 + 1)) & + (d_0 \times 1) \\
&= (d_2 \times 99) + (d_2 \times 1) + (d_1 \times 9) + (d_1 \times 1) + (d_0 \times 1) \\
&= (d_2 \times 99) + (d_1 \times 9) + (d_2 \times 1) + (d_1 \times 1) + (d_0 \times 1) \\
&= (d_2 \times 99) + (d_1 \times 9) + ((d_2 + d_1 + d_0) \times 1) \\
&= (d_2 \times 99) + (d_1 \times 9) + (d_2 + d_1 + d_0).
\end{aligned}
$$

Thus, $d_2d_1d_0$ can be expressed as a bunch of stuff that is definitely divisible by 9, plus the sum of its digits. So it is divisible by 9 if and only if the sum of its digits is divisible by 9.

$$
d_2d_1d_0 = \underbrace{(d_2 \times 99)}_{\text{divisible by 9}} + \underbrace{(d_1 \times 9)}_{\text{divisible by 9}} + \underbrace{(d_2 + d_1 + d_0)}_{\text{sum of digits}}.
$$

Of course, that deals only with numbers in the hundreds. But we could extend the argument to bigger numbers. For instance, what would you write for the number 61 237? And how would the argument go for the general 5-digit number

$$
d_4d_3d_2d_1d_0 = (d_4 \times 10\,000) + (d_3 \times 1000) + (d_2 \times 100) + (d_1 \times 10) + (d_0 \times 1)?
$$

Finally, why do you think I labelled the digits $d_2d_1d_0$, with d_0 as the units digit? In one sense, this doesn't matter—notation can be set up in any convenient way. But calling the units digit d_0 permits consistent labelling for bigger numbers: as just shown, $d_4d_3d_2d_1d_0$ can represent a number in the tens of thousands. Moreover, labelling the units digit d_0 links to powers of 10 in the base-10 system:

$$
\begin{array}{ccccc}
10^4 & 10^3 & 10^2 & 10^1 & 10^0 \\
10\,000 & 1000 & 100 & 10 & 1 \\
\text{tens of thousands} & \text{thousands} & \text{hundreds} & \text{tens} & \text{units}
\end{array}
$$

We can also extend to negative powers. And, because we are working in base 10, we can add a row of decimal representations (negative powers are discussed in Section 3.6):

10^0	10^{-1}	10^{-2}	10^{-3}	10^{-4}
1	$\frac{1}{10}$	$\frac{1}{100}$	$\frac{1}{1000}$	$\frac{1}{10\,000}$
1	0.1	0.01	0.001	0.0001
units	tenths	hundredths	thousandths	ten thousandths

This chapter will involve decimals in association with division. But, for now, I want to ask whether the dividing-by-9 result works similarly for other numbers. Does it? Or is there something special about 9s? What do you think? Clearly there is something special in that powers of 10 are 'nearly' multiples of 9. But it is also true that

Claim: a number is divisible by 3 if and only if the sum of its digits is divisible by 3.

Can you see why? It works in *some* cases because if a number is divisible by 9, it is also divisible by 3. But many numbers are divisible by 3 but not by 9. For instance, $738 - 3 = 735$ must have this property. Not much adjustment is needed, though, to apply a similar argument. Because 99 and 9 are both divisible by 3, a number (in the hundreds) is divisible by 3 if and only if the sum of its digits is divisible by 3:

$$735 = \underbrace{(7 \times 99)}_{\text{divisible by 9}} + \underbrace{(3 \times 9)}_{\text{divisible by 9}} + \underbrace{(7 + 3 + 5)}_{\underline{\text{not}} \text{ divisible by 9}},$$

$$735 = \underbrace{(7 \times 99)}_{\text{divisible by 3}} + \underbrace{(3 \times 9)}_{\text{divisible by 3}} + \underbrace{(7 + 3 + 5)}_{\text{divisible by 3}}.$$

In general,

$$d_2 d_1 d_0 = \underbrace{(d_2 \times 99)}_{\text{divisible by 3}} + \underbrace{(d_1 \times 9)}_{\text{divisible by 3}} + \underbrace{(d_2 + d_1 + d_0)}_{\text{sum of digits}}.$$

How does this work for a number in the tens of thousands, or the millions? And are there more divisors with similar properties to 9 and 3?

What would be the answers if we worked in base 5 instead of base 10? And have you thought much about the phrase *if and only if*?

5.3 If and only if

The phrase *if and only if* appeared several times in the previous section. You probably didn't notice, so you might want to look back and check. But maybe you remember that I both used and emphasized it in Section 1.7 when talking about Pythagoras' theorem. The phrase *if and only if* is heard only rarely outside mathematics, but in mathematics it crops up all the time because it captures an important type of logical relationship. In this case, it is used in a single statement capturing two claims:

Claim: A number is divisible by 9 *if* the sum of its digits is divisible by 9.

Claim: A number is divisible by 9 *only if* the sum of its digits is divisible by 9.

This sounds simple, but it's not. In everyday life, the word *if* is bandied about with all sorts of ambiguity. English speakers use it in logically inconsistent ways, but nobody notices because we all ignore the logic and interpret what people say according to what, in the circumstances, they probably mean. You might, therefore, wonder what I'm talking about—maybe to you these two statements seem to say the same thing. They don't say the same thing, but that can be hard to see when both are true. It's easier when two statements have the same structures but distinct *truth values*. For instance, one claim below is true and one is false. Which is which?

Claim: A number is divisible by 5 *if* its last digit is 5.

Claim: A number is divisible by 5 *only if* its last digit is 5.

The first is true: a number is certainly divisible by 5 if its last digit is 5 (think about 5, 25, 1005, etc.). The second is false: a number can be divisible by 5 without having last digit is 5. It could, instead, have last digit 0 (think about 10, 30, 610, etc.). So these two statements definitely don't say the same thing—if one is true and the other is false, there must be an important difference.

I find it helpful to think about this using Venn diagrams. Mathematicians say that the set of numbers with last digit 5 is a *proper subset* of the set of numbers that are divisible by 5. For those who prefer words, mathematicians also say that having last digit 5 is a *sufficient condition* for a number to be divisible by 5, but not a *necessary condition*. How does that relate to these diagrams?

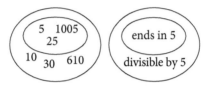

If you're still wondering what I'm talking about—perhaps these examples seem straightforward—try applying the same thinking to this statement:

You can have ice cream if you finish your vegetables.

We've all heard this kind of thing—if you spend time around children, you might say something like it every day. What is its Venn diagram?

Now, does the sentence correspond logically to what the person means? No, it does not. It says something about what happens if you finish your vegetables, but nothing about what happens if you don't. If you don't immediately see the problem, consider this conversation.

ADULT: You can have ice cream if you finish your vegetables.

CHILD: And what if I don't? Can I have it then too?

ADULT: Don't be cheeky—you know what I meant.

The child's response is cheeky precisely because it is logically astute—it's a literal interpretation of what was said, presumably in full knowledge that this is not what was meant. Mathematically speaking, when a person says

You can have ice cream if you finish your vegetables,

what they actually mean is

You can have ice cream only if you finish your vegetables,

or maybe

You can have ice cream if and only if you finish your vegetables.

But no one talks like that. And I'm not about to suggest that we start—people are not wrong to be sloppy with *if*, because everyday communication works perfectly well. My point is not even that mathematicians are fussier than most people, though they certainly are, at least about issues of logic. My point is that expected levels of precision can vary with context. Doubtless you use language differently when talking to your boss, to your friends, or to people who speak English as a second language. Mathematics is just another context: communicating well mathematically involves learning about when to get fussy.

5.4 Division and decimals

The material on dividing by 9 focused on exact divisibility. What happens when division is not exact? We can handle that using remainders or decimals. We'll start with the latter, but both ideas will appear in the coming sections.

Recall from Chapter 3 that

$$\frac{2}{7} = 0.285714\ldots \quad \text{and} \quad \frac{4}{11} = 0.363636\ldots.$$

Chapter 3 focused on fractions as ratios and as numbers that can be represented on a number line. Here we will focus on fractions as they relate to division, where we could read

$\frac{2}{7} = 0.285714\ldots$ as   'two divided by seven equals $0.285714\ldots$'.

When reading in this way, we tend to think of $2 \div 7$ as the calculation, and $0.285714\ldots$ as the answer. But mathematicians would tell you that while $0.285714\ldots$ (with the ellipsis) could be thought of as the answer, any finite portion of it, 0.285714 for instance, could not. The finite decimal might be a very good approximation. But, no matter how many digits we write, it is not exactly equal to $2 \div 7$. Why are they so confident about that, though? How do they know that the decimal doesn't terminate after, say, 25 digits?

To explore this we will start with $\frac{4}{11} = 0.363636\ldots$, which is simpler because its decimal form has a repeating pattern. The pattern means that $4 \div 11$ could be written as $0.\dot{3}\dot{6}$ or $0.\overline{36}$ (both indicate that the digits $3, 6$ repeat forever). But, again, this claim is not obviously true. It *seems* to be true, but six digits isn't many in a potentially infinite decimal expansion. What if this pattern goes on for a while and then changes?

In fact, the pattern does not change, and the way to see this is by long division. Yes, you read that correctly. Long division is good for something. For calculations it's now anachronistic—people can obtain numerical answers in as few seconds as it takes to press the relevant calculator buttons. But pressing buttons provides no insight. A calculator will tell you *that* the decimal form of $4 \div 11$ has a repeating pattern in its first few digits, but it won't confirm that this pattern goes on forever, and it certainly won't tell you *why* the pattern exists. For that, we need to take apart the process.

 I will describe long division with a focus on meaning. I'll also start with $40\,000 \div 11$ rather than $4 \div 11$, because that will show the developing pattern without involving decimals. Readers of different ages will have had different experiences of long division. Do bear with me if my notation doesn't quite match what you learned in school, but you perhaps learned to say and write something like the following. Eleven into 4 doesn't go. Eleven into 40 goes 3, so we put the 3 over the 0 of the 40, subtract $3 \times 11 = 33$ from 40, and calculate the remainder, which in this case is 7. Then 11 into 7 doesn't go, so we 'bring down' another 0 to make 70, and so on.

```
        3                        3                        3
   ┌──────────           ┌──────────           ┌──────────
11 │ 4  0  0  0  0   11  │ 4  0  0  0  0   11  │ 4  0  0  0  0
     3  3                  3  3                  3  3
                      ─────────────         ─────────────
                          7                      7  0
```

This can look a bit magical, as if the numerals can slide around in a way that's detached from their place-value meanings. That's not what's happening, but this approach does tend to obscure the reasoning, and I think it's easier if we use all the 0s and track place value explicitly. With that in mind, we'll start again.

We want to find out how many 11s there are in 40 000, and we'll start by working out what the thousands digit of the answer must be. Four-thousand 11s is 44 000, which is too big: 4000 11s won't fit. But 3000 11s is 33 000 (three 11s is 33, and 1000 times that is 33 000). So 3000 11s will fit, and the thousands digit of the answer is 3. This is recorded in the thousands place at the top. There is still room for a lot more 11s, of course—to find out how much room we subtract the 33 000 to see what's left:

	tenthousands	thousands	hundreds	tens	units			tenthousands	thousands	hundreds	tens	units
		3							3			
11	4	0	0	0	0		11	4	0	0	0	0
	3	3	0	0	0			3	3	0	0	0
								7	0	0	0	

We have 7000 left, and we'll next work out what the hundreds digits of the answer must be. Seven hundred 11s is 7700, which is too big. But 600 11s is 6600 (six 11s is 66, and 100 times that is 6600). So 600 11s will fit, and the hundreds digit of the answer is 6. This is recorded in the hundreds place at the top, and we subtract the 6600 to see what's left:

	ten thousands	thousands	hundreds	tens	units
		3			
11	4	0	0	0	0
	3	3	0	0	0

	ten thousands	thousands	hundreds	tens	units
		3			
11	4	0	0	0	0
	3	3	0	0	0
		7	0	0	0

	ten thousands	thousands	hundreds	tens	units
		3	6		
11	4	0	0	0	0
	3	3	0	0	0
		7	0	0	0
		6	6	0	0
			4	0	0

We have 400 left, and we can get down to units in two more steps. Thirty 11s is 330 (three 11s is 33, and 10 times that is 330). So the tens digit of the answer is 3, and we subtract 330 and have 70 left. Finally, six 11s is 66, so the units digit of the answer is 6, and we are left with a remainder of 4.

	ten thousands	thousands	hundreds	tens	units
		3	6		
11	4	0	0	0	0
	3	3	0	0	0
		7	0	0	0
		6	6	0	0
			4	0	0

	ten thousands	thousands	hundreds	tens	units
		3	6	3	
11	4	0	0	0	0
	3	3	0	0	0
		7	0	0	0
		6	6	0	0
			4	0	0
			3	3	0
				7	0

	ten thousands	thousands	hundreds	tens	units
		3	6	3	6
11	4	0	0	0	0
	3	3	0	0	0
		7	0	0	0
		6	6	0	0
			4	0	0
			3	3	0
				7	0
				6	6
					4

We've now established that 40 000 ÷ 11 is 3636 remainder 4. But why were we doing this, and what have we learned? We were doing it to see why the pattern in the answer repeats: 3 then 6 then 3 then 6. Can you see why? It's because the pattern of remainders in the calculation repeats: 7 then 4 then 7 then 4. This happens because the nearest we can get to 40 with elevens is $3 \times 11 = 33$, which puts a 3 in the answer and a remainder of 7 in the calculation (so we next work with a 7 followed by 0s). The nearest we can get to 70 with 11s is $6 \times 11 = 66$, which puts a 6 in the answer and a remainder of 4 in the calculation (so we next work with a 4 followed by 0s). You might want to pause here, read back a bit, and make sure you're convinced about this.

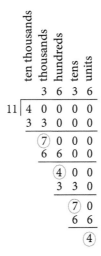

Now, how about $4 \div 11$? Perhaps you can see that if $40\,000 \div 11 = 3636$ plus a remainder, then $4 \div 11 = 0.3636$ plus a 'remainder' in a sort of decimal sense. Perhaps you are also convinced that the decimal repeats forever. It's worth nailing down how this works, though. To calculate $4 \div 11$, we ask how many 11s there are in 4. Not even one will fit. But, in the spirit of long division, we can think of 4 as 40 tenths. Dividing 40 tenths by 11 gives 11 lots of 3 tenths, totalling 33 tenths, with 7 tenths left over.

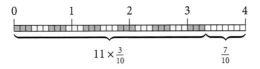

In the written calculation this means that we need a 3 in the tenths place of the answer, then to subtract 33 tenths or 3.3, leaving seven tenths or 0.7:

	units	tenths	hundredths	thousandths	ten thousandths
	0.	3			
11	4.	0	0	0	0
	3.	3	0	0	0
	0.	7	0	0	0

Now we can treat 0.7 as 70 hundredths. Dividing that by 11 gives 11 lots of 6 hundredths, totalling 66 hundredths, with 4 hundredths left over (a diagram for this wouldn't fit across the page—can you imagine one?). This means that we need a 6 in the hundredths place of the answer, then to subtract 66 hundredths or 0.66, leaving 4 hundredths or 0.04. Then we're back to asking how many 11s go into a number with a leading 4 followed by 0s. So the pattern, as before, repeats. Here are a couple more steps. How would you describe them?

	units	tenths	hundredths	thousandths	ten thousandths
	0.	3			
11	4.	0	0	0	0
	3.	3	0	0	0
	0.	7	0	0	0

	units	tenths	hundredths	thousandths	ten thousandths
	0.	3	6		
11	4.	0	0	0	0
	3.	3	0	0	0
	0.	7	0	0	0
	0.	6	6	0	0
	0.	0	4	0	0

	units	tenths	hundredths	thousandths	ten thousandths
	0.	3	6	3	
11	4.	0	0	0	0
	3.	3	0	0	0
	0.	7	0	0	0
	0.	6	6	0	0
	0.	0	4	0	0
	0.	0	3	3	0
	0.	0	0	7	0

This is what convinces mathematicians that the pattern in $4 \div 11 = 0.363636\ldots$ goes on forever. It's not seeing the first few digits on a calculator screen; it's understanding the process that leads to the repetition.

5.5 Decimals and rational numbers

The argument in the previous section did not use a calculator, but it did stick to the 'calculate' direction: we converted $4 \div 11$ to an 'answer' in decimal form. How about the other direction, converting decimals to fractions? Do you know how to do that? Probably you know some specific conversions: $0.5 = \frac{1}{2}$, and $0.333\ldots = \frac{1}{3}$, and so on. But perhaps you don't know how to convert decimals to fractions in general—this doesn't always appear in school mathematics. Rather delightfully, though, we can do it in some cases by adapting an argument from Section 3.7.

First, let's review what happens when we multiply by 10 or a power of 10. To multiply by 10 we 'add a zero'. For instance, $27 \times 10 = 270$. Alternatively, and more correctly, multiplying by 10 moves every digit one place to the left. This means that multiplying by 100 moves every digit two places to the left: $100 = 10 \times 10$, so multiplying by 100 is equivalent to multiplying by 10 twice. For instance,

$$27 \times 100 = 27 \times 10 \times 10 = 270 \times 10 = 2700.$$

thousands	hundreds	tens	units	
		2	7	
	2	7	0	$\longleftarrow$ one place to the left
2	7	0	0	$\longleftarrow$ two places to the left

How many places do the digits move if we multiply by a thousand? And if we multiply by a million? And why do the answers fit so neatly with the fact that one thousand $= 1000 = 10^3$ and one million $= 1\,000\,000 = 10^6$?

For a decimal example, consider 0.2, which is two tenths. Multiplying 0.2 by 10 gives 20 tenths, which is 2:

$$0.2 \times 10 = \tfrac{2}{10} \times 10 = \tfrac{20}{10} = 2.$$

For a more complex decimal example, consider 0.027. This is two hundredths and seven thousandths or, if you prefer, 27 thousandths.

Multiplying by 10 gives two tenths and seven hundredths, or 27 hundredths. Multiplying by 100 gives two units and seven tenths, or 27 tenths:

units	tenths	hundredths	thousandths	
0.	0	2	7	
0.	2	7	0	⟵ one place to the left
2.	7	0	0	⟵ two places to the left

In these examples, all the numbers have *terminating decimal expansions*, meaning that they can be written with a finite number of digits after the decimal point. But the same applies to nonterminating expansions, and we can exploit that to convert repeating decimals to fractions. Read this argument carefully—I think you'll like it:

$$\text{Let} \qquad x = 0.272727\ldots$$
$$\text{Then} \qquad 100x = 27.272727\ldots$$
$$\text{So} \qquad 100x - x = 27$$
$$\text{So} \qquad 99x = 27$$
$$\text{So} \qquad x = \frac{27}{99} = \frac{3}{11}.$$

This works because multiplying by 100 moves *all* the digits two places to the left, so everything lines up perfectly and subtracting x from $100x$ gives a whole number. You might also notice that in the last line I rewrote the fraction in *lowest terms*. This is an aesthetic choice. It is perfectly correct to conclude that $0.272727\ldots = 27/99$. But 27 and 99 have *common factor* 9. Dividing the numerator and denominator by this common factor gives an equivalent fraction that is 'nicer' in the sense that its components are smaller.

Now, for what other numbers would this process work? Yep, all numbers with two-digit repeating decimal expansions. You might like to try it with $0.363636\ldots$, reduce your answer so that the fraction is in lowest terms, and check that you get the expected familiar number.

Next, how about three-digit repeating blocks?

Let $x = 0.738738738\ldots$

Then $1000x = 738.738738738\ldots$

So $1000x - x = 738$

So $999x = 738$

So $x = \dfrac{738}{999}.$

Is the result in lowest terms? No. As we know, 738 is divisible by 9, so

$$0.738738738\ldots = \frac{738}{999} = \frac{82}{111}.$$

How about numbers that have repeating patterns but are not *proper fractions* because they are greater than 1? Here is an example:

Let $x = 12.3123123\ldots$

Then $1000x = 12\,312.3123123\ldots$

So $1000x - x = 12\,300$

So $999x = 12\,300$

So $x = \dfrac{12\,300}{999}.$

If you're not a fan of improper fractions—if you'd prefer that answer written as a whole number plus a proper fraction—we could lop 12 off at the beginning then add it back on at the end.

Let $x = 0.3123123\ldots$

Then $1000x = 312.3123123\ldots$

So $1000x - x = 312$

So $999x = 312$

So $x = \dfrac{312}{999}.$

Hence $12.3123123\ldots = 12 + \dfrac{312}{999}.$

If you can see these arguments as generic, you might notice something interesting. *Every* repeating decimal can be converted to a fraction. Did you know that? If not, you might pause for a moment to appreciate it.

Then ask yourself, *is the converse true*? Does every fraction have a repeating decimal expansion? To answer this, we first need to clarify the meaning of 'repeating'. The arguments in this section have used only numbers where the repeating block forms the entire number. But that's not necessary: the numbers 0.12343434... and 953.721111... are also considered repeating. So are the numbers 0.37500000... and 12.0000000.... Normally we'd write 0.375 and 12 and say that these have *terminating* decimal expansions. But, treated as infinite decimals, they end in repeating 0s. So they are considered repeating (albeit as degenerate cases).

With that in mind, what do you think? Does every fraction have a repeating decimal expansion? We can think about this using Venn diagrams. We have established that every repeating decimal represents a fraction or, more properly, a rational number. So we are in one of the following two situations. Maybe the rationals and the repeating decimals are exactly the same numbers, so that a number is rational *if and only if* its expansion is repeating. Or maybe there are extra rationals that do not have repeating expansions, so that a number is rational *if* its expansion is repeating, but not *only if* its expansion is repeating. What do you think? And, whatever your intuition says, could you justify it?

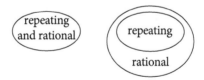

We'll explore this by returning to $2 \div 7$ and long division. For long division with single-digit numbers, I favour a compact horizontal representation. The reasoning is the same as before but, because the remainders are single digits, we can squeeze them in next to the 0s. This calculation starts with the question of how many 7s there are in 2. Not even one will fit, but we can think of 2 as 20 tenths. Twenty tenths is seven lots of two tenths with six tenths left over:

$$\begin{array}{r} 0.\,2 \\ 7\,\overline{\smash{\big)}\,2.\,0\ ^60\ 0\ 0\ 0\ 0} \end{array}$$

Then we can treat the six tenths as 60 one-hundredths. Sixty one-hundredths is seven lots of eight one-hundredths with four one-hundredths left over. Continuing generates the first six digits, 0.285714:

$$\begin{array}{r} 0.\,2\ 8\ 5\ 7\ 1\ 4 \\ 7\,\overline{\smash{\big)}\,2.\,0\ ^60\ ^40\ ^50\ ^10\ ^30\ ^20} \end{array}$$

And the next remainder is 2, putting us in a familiar situation: the number we want to divide by 7 is 2 followed by 0s. So the pattern starts to repeat:

$$\begin{array}{r} 0.\,2\ 8\ 5\ 7\ 1\ 4\ 2\ 8\ 5\ 7\ 1\ 4\ldots \\ 7\,\overline{\smash{\big)}\,2.\,0\ ^60\ ^40\ ^50\ ^10\ ^30\ ^20\ ^60\ ^40\ ^50\ ^10\ ^30\ ^20\ldots} \end{array}$$

For the question about rationals and decimals, we need to ask whether this is a coincidence. Is $\frac{2}{7}$ special? Or does something similar happen for every rational number? The key, again, is to think about the process. When we divide by 7, how many possible nonzero remainders are there? Just six: the remainder can only be 1, 2, 3, 4, 5, or 6. So, after at most six steps in the calculation, there must be a remainder that has come up before, and the decimal must repeat. The long division for $\frac{2}{7}$ cycles through all six possible remainders, giving a decimal expansion with repeating blocks of six digits, or *period* six. That doesn't always happen: when dividing by 11, there are ten possible remainders, but we saw that only two pop up when calculating the decimal expansion $\frac{4}{11} = 0.363636\ldots$, which therefore repeats with period two. But arguments like this constrain the period. When dividing by 73, only 72 remainders are possible, so the resulting decimal must repeat with period at most 72. And, when dividing by n, only $n-1$ remainders are possible, so the resulting decimal must repeat with period at most $n-1$.

There is a subtlety to consider, because a calculation could give a zero remainder. But that would also lead to a repeating pattern by giving a decimal expansion that ends in repeating 0s. For example, calculating $\frac{3}{8}$ gives

$$\begin{array}{r} 0.\,3\ 7\ 5\ 0\ 0\ 0\ 0\ldots \\ 8\overline{)\,3.\,0\ ^60\ ^40\ ^00\ ^00\ ^00\ ^00\ldots} \end{array}$$

Thus, every rational number has a repeating decimal expansion, and it is true that

Theorem: A number is rational *if and only if* its decimal expansion is repeating.

You've probably looked at loads of decimals. Did you know this about them? If not, you might want a bigger pause now. Every rational number has a repeating decimal expansion, and every repeating decimal represents a rational number. That's a pretty fundamental thing to understand about number representation.

5.6 Lowest terms

How does the decimal-to-rational conversion work for a number with a long period like $0.285714285714\ldots$, which we know is equal to $\frac{2}{7}$? Let's do it and see:

$$\begin{aligned} \text{Let} \qquad\qquad x &= \qquad 0.285714285714\ldots \\ \text{Then} \qquad 1\,000\,000x &= 285\,714.285714285714\ldots \\ \text{So} \qquad 1\,000\,000x - x &= 285\,714 \\ \text{So} \qquad\qquad 999\,999x &= 285\,714 \\ \text{So} \qquad\qquad\qquad x &= \frac{285\,714}{999\,999}. \end{aligned}$$

That all works. But it's not obvious that the final number is equal to $\frac{2}{7}$. It's about the right sort of size, but I can't see the equivalence, immediately or even with a bit of thought. I do know, though, that because this number is equal to $\frac{2}{7}$, it must be true that

$$\frac{285\,714}{999\,999} = \frac{2 \times \text{something}}{7 \times \text{something}},$$

where the somethings are both the same. I'd like to convince myself of this, preferably 'by hand' rather than by using a calculator. One approach is to consider the factors of the numerator and the denominator. The numerator is even so it has 2 as a factor. But *factoring* it—finding all of its factors—doesn't look easy. And I'm quite lazy so I'm not inclined to start there. The denominator, on the other hand, clearly has 9 as a factor, and dividing it by 9 will be easy. That gives a *factor tree* like this:

What about factoring 111 111? That looks harder, until you notice that the sum of its digits is a multiple of 3, so 111 111 must be a multiple of 3. We can find its corresponding factor by long division:

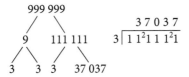

Now, it'd be great if the sum-of-digits thing worked for 37 037, but unfortunately it doesn't (check). So this calls for a bit of systematic thinking. The number 37 037 is not even so it is not divisible by 2 or by any other even number. It is not divisible by 3 because the sum of its digits is not divisible by 3. It is not divisible by 5 because it does not end in a 5 or a 0. Is it divisible by 7? We don't have any shortcuts for that, but long division shows that it is:

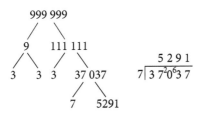

Now, 5291 is not divisible by 2 or by 3 or by 5, by similar reasoning. It's not divisible by 7 either—try the long division. What shall we try next?

Not 8 because that is even. And not 9 because a number can't be divisible by 9 without being divisible by 3 (make sure you understand why). And not 10 because 5291 does not end in a 0 (or, if you prefer, because 5291 is not even or is not divisible by 5). Do we need any of this reasoning, though? In fact, no. If 5291 had any of these factors, then 37 037 would too, and we already know that it doesn't. Again, make sure you understand that argument.

How about dividing 5291 by 11?

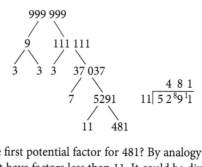

And what's the first potential factor for 481? By analogy with the earlier point, 481 can't have factors less than 11. It could be divisible by 11, because 5291 might have 11 as a repeated factor. Try and you'll see that it doesn't, though. It can't have 12 as a factor because it doesn't have 3 or 4. So the next thing to try is 13:

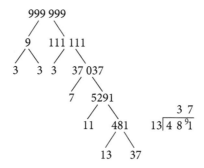

And that's it. We can't factorize further because all the numbers at the bottoms of the branches are *prime*, meaning that their only factors are themselves and 1. Hence, the *prime factorization* of 999 999 is

$$999\,999 = 3 \times 3 \times 3 \times 7 \times 11 \times 13 \times 37.$$

This helps with the idea that

$$\frac{285\,714}{999\,999} = \frac{2 \times \text{something}}{7 \times \text{something}},$$

because we can reorder the factors in 999 999 and write

$$\frac{285\,714}{999\,999} = \frac{2 \times \text{something}}{7 \times (3 \times 3 \times 3 \times 11 \times 13 \times 37)}.$$

Then, because we know that the fraction is equal to $\frac{2}{7}$, we must have

$$\frac{285\,714}{999\,999} = \frac{2 \times (3 \times 3 \times 3 \times 11 \times 13 \times 37)}{7 \times (3 \times 3 \times 3 \times 11 \times 13 \times 37)}.$$

However, I'm a bit nonplussed by that. I believe it—I'm confident of the reasoning so I know it must be true. But I can't 'see' that $285\,714 = 2 \times (3 \times 3 \times 3 \times 11 \times 13 \times 37)$. I could check with a calculator, but that seems defeatist. When drafting this chapter I stared at this for a while, then realized that I could use information from the factor tree, at least for the three most 'difficult' factors:

$$285\,714 = 2 \times (3 \times 3 \times 3 \times \overbrace{11 \times 13 \times 37}^{5291}).$$

Then, because $2 \times 3 \times 3 \times 3 = 6 \times 9 = 54$, it should be true that $285\,714 = 54 \times 5291$. That's just one calculation, and those who are comfortable with long multiplication might like to see it written like this:

$$
\begin{array}{r}
5\,2\,9\,1 \\
\times \quad\quad 5\,4 \\
\hline
2\,1\,1\,6\,4 \\
2\,6\,4\,5\,5\,0 \\
\hline
2\,8\,5\,7\,1\,4 \\
\hline
\end{array}
$$

If your long multiplication is rusty, you might like to relate that single calculation to the following two, which separate out the 50 and the 4 and show component calculations on different lines. On the left, for instance, $4 \times 1 = 4$, $4 \times 90 = 360$, and so on.

```
      5 2 9 1              5 2 9 1
  ×         4          ×        5 0
  ─────────────        ─────────────
            4                  5 0
        3 6 0              4 5 0 0
        8 0 0          1 0 0 0 0
    2 0 0 0 0      2 5 0 0 0 0
  ─────────────    ─────────────────
    2 1 1 6 4      2 6 4 5 5 0
```

Alternatively, you might prefer a grid representation that explicitly uses place value. The top splits 5291 into $5000 + 200 + 90 + 1$, and the left side splits 54 into $50 + 4$. Each box contains the product of the numbers at the top and the left. Adding these together gives the final total:

	5000	200	90	1
50	250 000	10 000	4500	50
4	20 000	800	360	4

```
  250 000
   20 000
   10 000
      800
    4 500
      360
       50
+       4
─────────
  285 714
```

In any case, I'm now satisfied that

$$\frac{285\,714}{999\,999} = \frac{2 \times (3 \times 3 \times 3 \times 11 \times 13 \times 37)}{7 \times (3 \times 3 \times 3 \times 11 \times 13 \times 37)} = \frac{2}{7}.$$

Not obvious, but true.

Here we've considered *prime factorization*, and you might know that every number has a *unique* prime factorization. When I began writing, I didn't know the prime factorization of 999 999, but I knew that it would have one and only one. Why is that true? What is special about primes? The answer is perhaps most easily seen by thinking about factorizations for simpler numbers. For instance, 12 can be factored in different ways:

$$12 = 2 \times 6, \qquad 12 = 4 \times 3.$$

So general factorizations are not unique. But underlying both of these is the same prime factorization:

$$12 = 2 \times 6 = 2 \times (2 \times 3);$$
$$12 = 4 \times 3 = (2 \times 2) \times 3.$$

Bracketing differently is possible because natural number multiplication is *associative* (see Section 2.10). And this can happen in more ways for numbers with more complex factor structures. For instance:

$$120 = 2 \times 60 = 2 \times (2 \times 2 \times 3 \times 5);$$
$$120 = 4 \times 30 = (2 \times 2) \times (2 \times 3 \times 5);$$
$$120 = 8 \times 15 = (2 \times 2 \times 2) \times (3 \times 5);$$
$$120 = 24 \times 5 = (2 \times 2 \times 2 \times 3) \times 5.$$

We can shuffle everything around to give other factorizations too, because multiplication is commutative. Here are a couple of examples:

$$120 = 3 \times 40 = 3 \times (2 \times 2 \times 2 \times 5);$$
$$120 = 6 \times 20 = (2 \times 3) \times (2 \times 2 \times 5).$$

What we can't do is split a number further than its prime factorization. We can't split a 2 or a 3 (or a 5 or a 7 or an 11 or a 13), because these and all primes are already split as far as they will go—they have no smaller integer factors (other than 1). This means that factorization is related to writing fractions in lowest terms, which will be handy in the next section. To conclude this one, you might like to work out how many ways are there to write 120 as a product of two positive integers, and what this has to do with its prime factorization.

5.7 Irrational numbers

We've now looked at various ways to represent numbers. Some, like $0.285714\ldots$, highlight magnitudes but are imprecise. Others, like $\frac{2}{7}$, are precise but obscure magnitudes. Still others, like $285714 / 999999$, are again precise but far from simple. They are nevertheless perfectly good representations.

But which numbers can we now represent in which forms? This returns us to the opening question about what numbers *are*. For instance, it would be reasonable to assume that every number can be represented as a fraction. Certainly any particular number has rationals really close by. For a classic example, a commonly cited rational approximation to the number π (pi) is 22/7:

$$22/7 = 3.14285714\ldots$$
$$\approx 3.14159265\ldots$$
$$= \pi.$$

Note that to read this aloud, we would say

> 'Twenty-two over seven is equal to 3.14285714…,
> which is approximately equal to 3.14159265…,
> which is equal to pi.'

I mention this because I've noticed that mathematics students are accustomed to seeing each new line linked by an '='. This makes some read the array as saying that 22/7 is equal to π, which is incorrect. Because 3.14285714… is only approximately equal to 3.14159265…, 22/7 is only approximately equal to π. That said, 22/7 is a good approximation—it's off by less than two one-thousandths. A less well-known but better approximation is 355/113, which is off by less than three ten-millionths:

$$355/113 = 3.14159292\ldots$$
$$\approx 3.14159265\ldots$$
$$= \pi.$$

Indeed, we can use decimals to get as close as we like:

$$\frac{314}{100} = 3.14, \qquad \frac{314\,159}{100\,000} = 3.14159, \qquad \frac{314\,159\,265}{100\,000\,000} = 3.14159265, \text{ etc.}$$

None of these are exact either. But such thinking makes many people believe that there will be a rational number exactly equal to π. After all, there are infinitely many choices of numerator and denominator so there are many, many rational numbers.

But what else do we know that's relevant? Think again about this:

Theorem: A number is rational *if and only if* its decimal expansion is repeating.

What does that mean about representing numbers as rationals? Not every number has a repeating expansion, so

Claim: *Not every number is rational.*

To find an *irrational number*, we just have to think up a nonrepeating decimal. As you might know, π is irrational, but it's pretty useless for this purpose because its lack of pattern means that I could present its first million digits without demonstrating that it doesn't repeat. Fortunately, though, it's not hard to construct a number that does have a clear but nonrepeating pattern. Here's one:

$$0.101001000100001000001\ldots$$

Look carefully—what is the pattern? Are you convinced that this could be extended forever without repeating? And could you vary it to construct more nonrepeating decimals? The lack of repeat means that this number and others like it cannot be represented in rational form—no combination of numerator and denominator would work. If you've reached the ends of other chapters you might recognize this as a *nonexistence* claim. We've demonstrated—convincingly I hope, if not formally—that rationals have repeating decimal expansions. This number does not have a repeating expansion, so it cannot be rational.

And this result—that some numbers are irrational—relates to Section 3.2. There, I observed that shifting from integers to rationals shifts our attention from numbers as discrete objects towards the number line as a continuum.

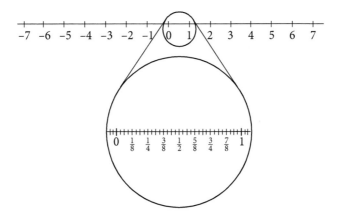

Now, however, we can observe that the rationals *do not fill the number line*. Although there are infinitely many, they leave lots and lots of 'gaps', one at every location where an irrational is needed.

So decimal expansions teach us about numbers. But for me it feels a bit unsatisfying to write irrationals only as decimals. I think it's nicer to represent numbers with compact symbols, like 'π'. I'd really like to explain how mathematicians know that π is irrational, but sadly that's beyond the scope of this book. However, I can provide a classic proof that $\sqrt{3}$ is irrational, which shows that some irrationals can be represented in compact forms.[1]

First, a reminder about square roots. As you probably know, the square root of n is a number that, when squared, gives n. Some integer examples are

$$\sqrt{1} = 1 \text{ because } 1^2 = 1 \times 1 = 1;$$
$$\sqrt{4} = 2 \text{ because } 2^2 = 2 \times 2 = 4;$$
$$\sqrt{9} = 3 \text{ because } 3^2 = 3 \times 3 = 9;$$
$$\sqrt{16} = 4 \text{ because } 4^2 = 4 \times 4 = 16, \text{ etc.}$$

The number $\sqrt{3}$ will not be an integer—think about where it would fit in this list. To prove that it is not only not whole but also not rational, it is useful to observe a link between squaring and prime factorization. The square of a number will have the same prime factors as that number, just twice as many of each. For instance,

$6 = 2 \times 3$ so $6^2 = (2 \times 3)^2 = 2 \times 3 \times 2 \times 3.$
$20 = 2 \times 2 \times 5$ so $20^2 = (2 \times 2 \times 5)^2 = 2 \times 2 \times 5 \times 2 \times 2 \times 5.$

This means that squaring cannot generate prime factors that are not in the original number. For n^2 to have 3 as a prime factor, n must have 3 as a prime factor. And that is key to a proof that $\sqrt{3}$ is irrational, which appears below. This proof is somewhat long, but don't be put off. Take it one line at a time, asking whether you understand each deduction. If you get stuck, read on anyway—sometimes things make more sense when you see how an argument fits together. It might also help to know that this is a *proof by contradiction*: it starts by assuming the opposite of what we expect, establishes that this leads to an impossible situation, and

[1] Those with mathematical backgrounds might wonder why I'm talking about $\sqrt{3}$ rather than $\sqrt{2}$. The reason is that people always do $\sqrt{2}$ and I fancied a change.

thereby concludes that what we expect must be true. Have a good bash at this—it's a classic and, if you've got this far, it's worth a bit of your time.

Claim: $\sqrt{3}$ is irrational.

Proof: Suppose to the contrary that $\sqrt{3}$ is *rational*.

Then we can write $\sqrt{3} = p/q$, where p and q are integers.

In fact, by appropriate division, we can write $\sqrt{3} = p/q$ where the fraction is in lowest terms, so p and q have no common factors.

Now $\sqrt{3} = p/q$ implies that $3 = p^2/q^2$ so $3q^2 = p^2$.

And $3q^2 = p^2$ means that 3 is a factor of p^2.

Because 3 is prime, this means that 3 must also be a factor of p.

Say $p = 3k$ where k is an integer.

Now $p = 3k$ implies that $3q^2 = 9k^2$, so $q^2 = 3k^2$.

And $q^2 = 3k^2$ means that 3 is a factor of q^2.

But then 3 must also be a factor of q.

So p and q have common factor 3.

But this contradicts the assumption that p and q have no common factors.

So the assumption leads to a contradiction: if $\sqrt{3}$ is rational, p and q have common factors and don't have common factors.

So $\sqrt{3}$ cannot be rational.

So $\sqrt{3}$ is irrational.

If you understood that, great, and you might like to think about how you would explain it to someone who didn't. If you didn't understand it, don't worry—it's probably the most challenging argument in the book, and you might find that you do understand it if you read it again in five minutes, or a month. Read on for now; the remaining sections draw together familiar ideas.

5.8 How many rationals and irrationals?

How many rational numbers are there? And how many irrationals? Infinitely many of each, obviously, so in one sense these questions are silly.

But they also encompass some profound mathematics, because it turns out to be meaningful to say that there are *more* irrationals than rationals. That surprises many people, for two reasons. First, because it's easy to list rationals and harder to think about irrationals. Second, because people have often thought of infinity as just one thing—it has never occurred to them that 'infinities' could be different 'sizes'. We'll work up to that idea.

I observed earlier that infinity, denoted by the symbol '∞', does not behave like a number. If it did, we would want to say that $1 \times \infty = 2 \times \infty$, so that dividing by ∞ would give $1 = 2$ and thus 'break' arithmetic. It would break arithmetic for addition and subtraction, too. For instance, if we take one away from a *finite* set of objects, we're left with fewer. In the following diagram, $7 - 1 = 6$, and $6 < 7$.

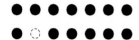

That's not true for infinite sets. Subtracting one from an infinite set of objects leaves infinitely many. So $\infty - 1 = \infty$; we don't have $\infty - 1 < \infty$:

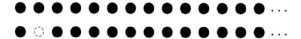

Similarly, $\infty - 2 = \infty$, and $\infty - 10 = \infty$, and $\infty - 1\,000\,000 = \infty$. Indeed, we could take away fully half of the objects and *still* have infinitely many. So we'd want to write $\infty - \infty = \infty$:

And it gets worse. Infinite arithmetic is inconsistent not only with finite arithmetic, but also with itself. If we took away all the objects, we'd have none left, meaning that we'd want to write $\infty - \infty = 0$:

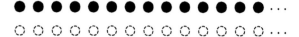

So treating ∞ as if it were a number leads to saying that $\infty - \infty = \infty$ and that $\infty - \infty = 0$. This makes the whole business look intractable.

But mathematicians are persistent—they don't give up when faced with difficulties. And they are flexible—they can accept that arithmetic doesn't work in the *usual* way for infinite sets, while maintaining interest in what *would* work. In this case, a satisfactory way to resolve the problem is to extend the concept of counting. Normally we think about counting finite sets. We point at the objects one at a time while reciting the count sequence '$1, 2, 3, \ldots$'. The last number we say is the number of objects or the *cardinality* of the set. Formally, counting involves a *one-to-one correspondence* between the objects and the first few natural numbers (which are sometimes called the 'counting numbers'):

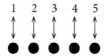

For an infinite set there will be no 'last' number, but the correspondence idea still works. For instance, we can 'count' the positive even numbers using the obvious correspondence with the natural numbers:

We never run out of either natural or even numbers, and the correspondence works forever. So it makes sense to say that these sets have the same cardinality, although that cardinality is infinite and the lack of a last number means that it doesn't have a pre-existing name. To avoid the ambiguous symbol '∞', mathematicians call it $\aleph_0$ ('aleph nought').

Now, it's fine if you don't like this idea. Maybe you'd prefer to say that there are half as many positive even numbers as natural numbers. But then you'd be back in the situation where infinite arithmetic creates inconsistencies. Those inconsistencies are mitigated by the correspondence approach, so that's the one mathematicians use.[2] Specifically, if there is a

[2] I'm simplifying here. There are other ways to formalize notions of infinity while maintaining consistency, and these constitute active areas of mathematical research. But this approach is considered standard and is taught to undergraduates.

one-to-one correspondence between the natural numbers and a set, that set is said to be *countably infinite* (or sometimes just *countable*).

You can learn more by looking up 'cardinal arithmetic'. But right now you might be wondering how there could ever fail to be correspondence, because surely an infinite list is long enough to contain all possible objects. In fact, though, *it isn't*. Understanding why is a big conceptual challenge, because for most people it involves loosening some previously unexamined but very entrenched intuitions. But we'll have a crack at it in the remainder of this section. Even those who've got this far might find this mind-bending, so remember that you can always skip to the next (and final) section and come back later.

First, the set of all integers is countable. Although in an everyday sense we might want to say that there are 'more' integers than naturals, in the correspondence sense the two sets have the same cardinality. We can demonstrate this by setting up a correspondence, which must go on forever and miss nothing out. Here is one way to do it:

$$
\begin{array}{cccccccc}
1 & 2 & 3 & 4 & 5 & 6 & 7 & 8 \dots \\
\updownarrow & \updownarrow & \updownarrow & \updownarrow & \updownarrow & \updownarrow & \updownarrow & \updownarrow \\
0 & 1 & -1 & 2 & -2 & 3 & -3 & 4 \dots
\end{array}
$$

The arrow below relates this to the integers' usual order.

Even better, *the rationals are countable too*. This is very far from obvious, I think. There really are a lot of rationals, and the choosing-an-order problem is much worse. As I observed in Section 3.2, rationals can't be counted in order from left to right: there is no 'next' rational number after $\frac{1}{2}$. Nevertheless, a lovely visual argument provides an appropriate one-to-one correspondence. The first step is to imagine all the rationals in an infinite two-dimensional array where the first row lists each integer, the second row lists each integer over 2, and so on:

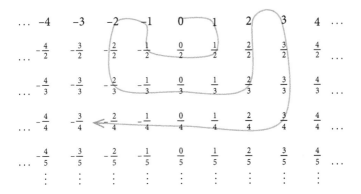

$$\ldots \quad -4 \quad -3 \quad -2 \quad -1 \quad 0 \quad 1 \quad 2 \quad 3 \quad 4 \quad \ldots$$

$$\ldots \quad -\frac{4}{2} \quad -\frac{3}{2} \quad -\frac{2}{2} \quad -\frac{1}{2} \quad \frac{0}{2} \quad \frac{1}{2} \quad \frac{2}{2} \quad \frac{3}{2} \quad \frac{4}{2} \quad \ldots$$

Then follow the arrow in the following diagram. Extended forever, this will hit every rational number, collecting them in an ordered list:

The arrow does not quite give a one-to-one correspondence because, for instance, $\frac{1}{2} = \frac{2}{4} = \frac{3}{6}$, and we don't want to 'count' this number multiple times. But that's easily dealt with. We can follow the arrow, but list only numbers that haven't come up before. Doing so leads to a correspondence that starts as in the following (check), and shows that the rationals are countable: they too have cardinality $\aleph_0$.

$$
\begin{array}{cccccccc}
1 & 2 & 3 & 4 & 5 & 6 & 7 & 8\ldots \\
\updownarrow & \updownarrow & \updownarrow & \updownarrow & \updownarrow & \updownarrow & \updownarrow & \updownarrow \\
0 & 1 & \tfrac{1}{2} & -\tfrac{1}{2} & -1 & -2 & -\tfrac{2}{3} & -\tfrac{1}{3}\ldots
\end{array}
$$

For many people this idea causes serious intellectual wobbles. They feel strongly that there are a lot more rationals than naturals, and that this therefore just cannot be right. If that's happening to you, it's perfectly normal. And we've been here before. At the end of Chapter 3 I observed that human intuitions tend to be based on small finite numbers.

But we're not working with those now, so the intuitions need loosening. What mathematicians do is loosen them in favour of more abstract notions about systemic consistency. Instead of thinking

'I must make this match my experience of everyday objects,'

they think

'Okay, clearly this does *not* match my experience of everyday objects, but perhaps we can still think about it in a way that's internally consistent.'

That frees them to explore various options, and the option that won out for dealing with infinite sets is the *Cantorian* one given here, named after Georg Cantor. It took a while, though—intuition loosening isn't easy for anyone. If your brain hurts, try putting this book down for a week. Brains are amazing self-organizers, and I almost guarantee that by then you'll find the idea more palatable.

If, however, you're still with me, here is the cool bit. The rationals are countable but the irrationals are *not*. While there is a one-to-one correspondence between the naturals and the rationals, there is no such correspondence between the naturals and the irrationals: there are so many irrationals that they cannot be arranged into an ordered list, even an infinite one. This is what it means to say that there are 'more' irrationals than rationals: the cardinality of the set of irrationals is greater than $\aleph_0$; it's a 'bigger' infinity. This can be demonstrated using another classic contradiction argument, which shows that any possible correspondence must miss out some irrationals. Here is that argument.

Consider all the rational and irrational numbers together. These fill the number line, and are known collectively as the *real numbers*. Think about the set of real numbers between 0 and 1 (for now), and imagine these written as decimals. Then suppose for contradiction that this set *is* countable, so there exists a one-to-one correspondence between this set and the natural numbers. The start of the correspondence would look like the following list, where each suffixed letter is a digit (we'd soon run out of letters, but you get the idea). As with other ideas about infinity, you should not imagine constructing this list as a process that happens in time. Instead, imagine that the entire infinite list already exists—someone claims to have a correspondence that works, and that includes every number between 0 and 1.

natural   corresponding decimal

$1 \longleftrightarrow 0.\ a_1\ a_2\ a_3\ a_4\ a_5\ a_6\ a_7\ a_8\ \ldots$

$2 \longleftrightarrow 0.\ b_1\ b_2\ b_3\ b_4\ b_5\ b_6\ b_7\ b_8\ \ldots$

$3 \longleftrightarrow 0.\ c_1\ c_2\ c_3\ c_4\ c_5\ c_6\ c_7\ c_8\ \ldots$

$4 \longleftrightarrow 0.\ d_1\ d_2\ d_3\ d_4\ d_5\ d_6\ d_7\ d_8\ \ldots$

$\vdots$

We will derive the contradiction by constructing a new number x that is between 0 and 1 but that is definitely *not* on the list (so the person must be wrong). We can ensure that x differs from the first listed number by ensuring that its first digit differs. For instance, if the first number is $0.6a_2a_3a_4\ldots$, we can set $x = 0.7x_2x_3x_4\ldots$ (switching the 6 for a 7). How can we ensure that x differs from the second listed number? We could work with its first digit again, but that's not a great strategy because we want x to differ from all the listed numbers, and we only have ten digits to play with. Instead, let's make x differ from the second listed number in the *second* digit. And we can make it differ from the third number in the third digit, from the fourth number in the fourth digit, and so on. Continuing like this gives a *diagonalization* argument, as represented in the following. We can make x differ from every listed number by making it differ from the nth number in the nth digit:[3]

$1 \longleftrightarrow 0.\ a_1\ a_2\ a_3\ a_4\ a_5\ a_6\ a_7\ a_8\ \ldots$

$2 \longleftrightarrow 0.\ b_1\ b_2\ b_3\ b_4\ b_5\ b_6\ b_7\ b_8\ \ldots$

$3 \longleftrightarrow 0.\ c_1\ c_2\ c_3\ c_4\ c_5\ c_6\ c_7\ c_8\ \ldots$

$4 \longleftrightarrow 0.\ d_1\ d_2\ d_3\ d_4\ d_5\ d_6\ d_7\ d_8\ \ldots$

$\vdots$

And there is the contradiction. The person who claimed to know how to construct this list was wrong. Although the list is infinitely long, it doesn't contain all the numbers between 0 and 1 because it doesn't include the one we just constructed.[4] And we could always perform that

[3] We should avoid introducing 0s or 9s because that might introduce duplicates: for instance, $0.1999\ldots = 0.2000\ldots$. But that's easy because there are plenty of digits to choose from.

[4] If you're thinking that we could add x to the end of the list, we couldn't. There is no end, and every natural number already has an assigned corresponding decimal.

construction. So the set of real numbers between 0 and 1 is not countable; its cardinality is bigger than $\aleph_0$. This means that the set of all real numbers is not countable. Thus, because the rationals *are* countable, the irrationals must be uncountable. There are more irrationals than rationals.

5.9 Number systems

As promised, a main theme of this chapter has been number systems: we've considered relationships between division, place value, rational numbers, and decimals. The preceding section explored the idea of systems more explicitly, considering inconsistencies that can arise when dealing with infinite sets. In this final section we'll pick up this consistency idea and relate it to more familiar finite arithmetic, tying up loose ends from across the book.

First, let's return to an idea from Section 3.6. There, I used the following diagram to explain why $2^0 = 1$, $2^{-1} = \frac{1}{2}$, and so on.

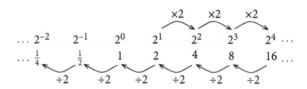

I also promised that we would look at this from a consistency perspective, which we'll do now. The expression 2^0 doesn't have an obvious meaning: we can't 'multiply 2 by itself zero times'. That being the case, we could give up and decide that 2^0 has no meaning. Or—and obviously this is the option that mathematicians took—we could work out whether it's possible to give it a meaning without breaking the system. Can we define 2^0 so that the relevant algebra still works? One algebraic rule is that $x^m x^n = x^{m+n}$. For this to work with 2^0, it should be true that

$$2^0 2^1 = 2^{0+1} = 2^1, \text{ that is, } 2^0 2^1 = 2^1.$$

But if 2^0 times 2^1 is 2^1, then 2^0 must be 1. So we should define 2^0 to be 1.

Similarly we can get to negative powers. Given that $2^0 = 1$, we can deduce that

$$2^{-1}2^1 = 2^{-1+1} = 2^0 = 1, \text{ that is, } 2^{-1}2^1 = 1.$$

But if 2^{-1} times 2^1 is 1, then 2^{-1} must be $\frac{1}{2}$. So we should define 2^{-1} to be $\frac{1}{2}$. Similar arguments show that we should define $2^{-2} = \frac{1}{4}$ and so on. Can you work out how?

We can generalize further, too, to fractional powers. For instance, what is $2^{\frac{1}{2}}$? This, again, is not obvious: we can't 'multiply 2 by itself half a time'. But if the algebraic rule $x^m x^n = x^{m+n}$ is to hold, it must be true that

$$2^{\frac{1}{2}}2^{\frac{1}{2}} = 2^{\frac{1}{2}+\frac{1}{2}} = 2^1, \text{ that is, } (2^{\frac{1}{2}})^2 = 2.$$

And if squaring $2^{\frac{1}{2}}$ gives 2, then $2^{\frac{1}{2}}$ must be $\sqrt{2}$. So we should define $2^{\frac{1}{2}}$ to be $\sqrt{2}$.

Now, it's possible that this talk about *defining* bothers you. Maybe you think that $2^0 = 1$ is just a fact, of the kind that appears in textbooks. But there was a time before textbooks, and a time before people thought anything at all about powers of 2. Historically, mathematicians did have to make a decision about this. Perhaps that's not what bothers you, though. Perhaps you feel that $2^0 = 1$ in a fundamental, law-of-the-universe sense. In that, I'm inclined to agree with you. In unreflective moments I'm a naive Platonist—my natural feeling is that 2^0 just is 1 and that it's nice that mathematicians have been clever enough to discover this. But I can also think in a more reflective and formal way about systemic consistency: 2^0 must be 1 in order to keep the systems working.

Next, a question about explanations. Which explanation do you like better: the diagram with the arrows or the argument about algebraic consistency? I prefer the arrows—I like intuitive, visual explanations. But the algebraic argument is better articulated and more mathematically formal. The real trick, in fact, is to see that both explanations capture the same mathematical ideas. I could try to explain how, but doing so would take a page or two and would probably be less effective than your just thinking about it. Do give it a go.

Finally, we'll link this right back to whole number arithmetic. Imagine yourself at a point in history when there was no formal education— no questions about 2^0 for you. But maybe you owned some things, and maybe you sometimes traded them. So, to keep track, you needed rudimentary counting. Now, how would you have felt about negative numbers? Probably you'd have found the idea peculiar. Perhaps you'd have thought it preposterous—no one, after all, has minus 3 sheep. Living in the 21st century, though, you have lots of experience with negative numbers. You can speak confidently about subzero temperatures or your overdraft or a sensible way of numbering a building's basement floors.[5] The language can sound a bit unnatural—I would say that I'm overdrawn by £50, not that I 'have minus £50'. But I can say 'It was minus 5 degrees last night' and know what that means, and I can relate all of these things to number-line scales:

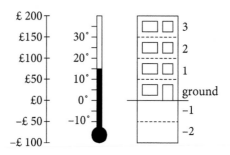

The scales don't do arithmetic, though. They leave open the question of what, for instance, $2 \times (-3)$ should be, or $(-2) \times (-3)$. But we can fix that. Just as we want operations with powers to maintain algebraic consistency, so we want operations with negative numbers to maintain arithmetic consistency. And again we can think about this using diagrams or arithmetic rules. I'll use some of each—you might like to think about alternatives.

Adding and subtracting can be represented on a number line. Adding 1 is moving one step right, subtracting 1 is moving one step left, and we can keep going to the left of 0 to label the negative numbers. Adding 3 is moving three steps right, and subtracting 3 is moving three steps left;

---

[5] Societies don't agree about the overground floors, of course—in the UK the first floor is the one above the ground floor, but that's not true everywhere.

addition and subtraction are *inverse operations*. This means that taking away –3 is *removing* a three-steps-left move, so subtracting –3 is the same as adding 3:

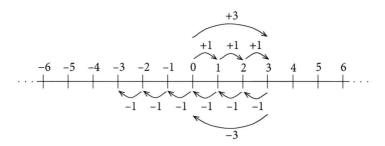

How about multiplication? Thinking informally, 2 × 3 is adding three twice, and 2 × (–3) is subtracting three twice.

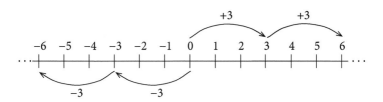

To formalize the latter we can ensure consistency with the distributive law,

$$(a \times b) + (a \times c) = a \times (b + c).$$

This law means that

$$(2 \times 3) + (2 \times (-3)) = 2 \times (3 + (-3))$$
$$= 2 \times 0$$
$$= 0,$$

So $\quad 2 \times (-3) = -(2 \times 3)$ (subtracting 2 × 3 from both sides)
$$= -6.$$

For $(-2) \times (-3)$ the argument uses this result and is basically the same:

$$(-2 \times 3) + (-2 \times (-3)) = -2 \times (3 + (-3))$$
$$= -2 \times 0$$
$$= 0,$$

So $\qquad -2 \times (-3) = -(-2 \times 3)$ (subtracting -2×3 from both sides)
$$= -(-6)$$
$$= 6.$$

Again this is about consistency. The equation $(-2) \times (-3) = 6$ is valid not because some authority woke up one day and decided that it should be. It's valid because multiplication of negative numbers must be this way to keep the distributive law working.

This, I hope, ties up some loose ends about arithmetic. But it opens up a lot more questions. For instance, $2^2 = 4$ and, by the reasoning we've just discussed, $(-2)^2 = (-2) \times (-2) = 4$. So 4 has two different square roots. What about -4? That doesn't seem to have any square roots: there is no number which, multiplied by itself, gives -4. But, just as it's possible to extend the positive number system to accommodate negative numbers, it's possible to extend again to accommodate square roots of negative numbers. Perhaps that's an idea for another book.[6]

5.10 Review

This chapter took a tour upward from base-10 representation through dividing by 9, decimals, long division, converting repeating decimals to fractions, prime factorization, irrational numbers, infinite sets, and countability; we arrived eventually at consistency within number systems.

On the way we thought about mathematical processes and structures. For the opening material on dividing by 9, considering the process of adding 9s gave some insight, but considering the base-10 structure of a number gave more. For the later material on decimal expansions, it was the other way around: the structure of a repeating expansion was visible

[6] Not everyone likes cliffhangers. If you want to learn about this, look up 'complex numbers'.

immediately, but understanding why it occurs required the process of division. In both cases, we used mathematical ideas to address specific questions. We considered place value not for its own sake, but to understand the division-by-9 phenomenon. We considered long division not for its own sake, but to understand why certain decimal expansions repeat. We also, as everywhere in this book, focused on meaning rather than repetitive calculation. I could sit around for ages converting repeating decimals to fractions, but I don't want to. I've seen enough to be convinced that it's always possible, and *that's* what's interesting—the general result that every repeating decimal expansion represents a rational number.

This chapter also contained a section about the phrase *if and only if*, and we used its ideas when relating rational and irrational numbers to decimal expansions. But the issues are ubiquitous in mathematics: mathematicians are very careful with the word *if*, and with broader logical reasoning. Logical reasoning is intimately tied to mathematical proof, as seen here in proofs by contradiction: in a proof by contradiction we say '*if* this were true, then a contradiction would arise'. And it guides the development of mathematics in the sense that consistency is paramount. The frontiers of mathematics are a long way from what's considered here, but mathematicians want every new abstract idea to fit in a consistent way with those we already have.

And now here we are at the end of the main content. I hope that you have learned a lot or made some new connections or perhaps just enjoyed seeing ideas linked together. In the Conclusion I will offer a few broader comments about mathematical learning.

Conclusion

This book is about mathematics, which means that it's about a lot of things.

First, it's about structure and pattern. In some places that's obvious: the chapter on shapes is all about patterns. But the content on numbers is about patterns, too. There is very little on individual numbers, because individual numbers are not that interesting. What's interesting is their collective regularity, their organization into structured number systems. Structure and pattern are the essence of mathematics, and attention to them is a real mathematical proclivity. If you know a child who likes to sort and organize things by their properties and symmetries, do encourage that.

Second, this book is about representations—about using representations to develop insight, and about their strengths and limitations for conveying mathematical ideas. I'm fond of diagrams—I like the way they capture lots of things at once. But I've also discussed the value of symbolic notation for brevity and for facilitating calculation, and the fact that mathematical language is precise but not rigid. Mathematicians combine symbols and words differently for different audiences. It is really only logicians who insist on symbolic grammar in the way that I might insist on where the apostrophes go. Mathematical communication is 'formal', but it is flexible too.

Third, the book is about arguments and evidence. Mathematics is a deductive science: it operates by rigorous logical reasoning. Mathematical proofs can have complex logical structures—proofs by induction or contradiction are only the tip of the iceberg. But all are based on simple deductive logic: if A is true, and A implies B, then B is true. Simple, that is, but not easy. The word 'if' causes no end of bother—it takes a lot of intellectual discipline to think about logic and avoid being swayed by context. And that, of course, is why evidence in the form of examples and diagrams is valuable. Reasoning is hard, and getting good ideas is even harder. Experienced mathematical thinkers use whatever is to hand to support their work.

Most importantly, this book is about the capacity of ordinary human beings to understand and enjoy mathematics. I didn't invent these mathematical ideas—they were created over thousands of years by mathematicians across the world. But I really like them. If you like them too, then I hope that you feel good about that, especially if you initially lacked confidence. The mathematics at the ends of the chapters would typically be introduced at university level. If you made good progress toward that—even if you didn't get to the end or digest everything—then maybe you're more mathematically capable than you thought.

Nevertheless, I hope that this book has left you with questions. Some might be about mathematics that you'd like to learn. But some might be about mathematics as an enterprise, and in the rest of this conclusion I'll address four of those:

- Why didn't my teachers explain it like that?
- What is it all for?
- What do mathematicians do?
- What shall I read next?

Why didn't my teachers explain it like that?

If you enjoyed this book and gained some insight, you might now feel both pleased with yourself and annoyed with your teachers. You might think,

Why didn't my teachers explain it like that?

and, possibly,

If they had, I'd have understood mathematics and enjoyed it!

These are natural reactions. But, before thinking too far along these lines, I suggest asking yourself a couple of questions. First, is it possible that your teachers *did* explain in these or closely related ways, but that you weren't listening? That you were distracted that week, or perhaps that year, by the things that routinely distract children and teenagers? Or that you expected teachers to 'tell you what to do', and were not open to understanding conceptual relationships? Second, is it possible that you are now a better thinker? If you are a moderately successful adult with the usual array of responsibilities, you'll have developed considerable intellectual maturity. You have more knowledge to which you can attach new information. And probably your logical reasoning has improved, so you are better able to do the attaching. So maybe your teachers were not at fault. It could be that I'm an unusually brilliant explainer. But it could be that I'm not. Maybe you're just a better learner.

My view is that the vast majority of teachers are extremely dedicated and do excellent work in a challenging environment. I have done only a little whole-day teaching, and only to adults, not to 14-year-olds. It is very, very demanding to be switched on for hours at a time, monitoring a room of individuals with disparate levels of understanding and engagement, responding in a positive and constructive way to unpredicted questions, and maintaining a sense of fairness and intellectual safety. And that's before you consider that teachers are often the first to identify and support children with serious problems, and that they have to change their teaching every year or two in response to authority-imposed policy and curriculum changes. I don't think I'd be a good schoolteacher—I think I'd be exhausted within a week—and it worries me that we can be so ready to criticize the people who are vital to our children's intellectual development. I think that can discourage people from becoming teachers when we need outstanding mathematical thinkers to inspire the next generation.

Of course, I also think that we can all be inspiring. If fewer adults said, 'Oh, I'm no good at maths' and more said, 'Oh yeah, maths is really interesting—I'd like to learn more', I think that would be great. It would help to create a positive buzz around the subject and to support those charged with our children's learning.

What is it all for?

If you are practically inclined, you might be willing to speak positively about mathematics but struggling to see the point. You might have spent much of the book thinking,

Okay, that's all very nice, but what is it *for*?

My answer to that would be, *How long have you got?* Mathematics underpins the contemporary world. It runs internet searches and encrypts bank details. It forecasts the weather and the price of wheat. It designs train timetables and schedules air crews. It processes medical images and informs decisions about how to spend taxes on healthcare in a way that maximizes benefit to the largest number of people. It prices your mortgage, your loans, your pension, your bus pass. It delivers the right amount of milk to your supermarket and to all the other supermarkets in the country. And so on.

If you're inclined to think that mathematics does a lot of these things imperfectly, maybe take a moment to think about that. In developed countries, it takes a pretty extraordinary social or meteorological event for the supermarkets to run out of milk. And they don't do it by having lots spare—no one wants to run massive refrigerated storage spaces or to lose money through waste. So someone must decide how to distribute all that milk, in all its different types and containers, so that your store gets the milk you want without a profligate waste of driver time and fuel. This is like the optimization problems in Chapter 4, except that it makes those look laughably trivial. If your immediate thought is that computers do all this stuff, think about that too. What calculations do the computers do? Who programs them to do those calculations, and how do those people evaluate the programs and improve them? It certainly can be annoying when we don't get these things perfectly right. But really it is astonishing that we can do them at all. For people interested in what mathematics is for, there are many, many answers.

To be honest, though, I don't care. I'm happy that mathematics is extraordinarily useful, but that's not why I like it. The pleasure it brings me is nothing to do with its utility—it's more akin to the pleasure I get from listening to music or viewing art or watching a great play or movie or TV show. I might learn something about myself or the world through these things, but that's not why I engage with them. I just love the experience. Maybe mathematics is not your thing and never will be, but I hope that you might now have a better sense of why some people think it's so great.

What do mathematicians do?

I definitely think that mathematics is great, but I also remarked in Chapter 2 that I'm not a mathematician. That probably sounded weird. I have both undergraduate and master's degrees in pure mathematics, and by most people's standards that's a lot. But, after that, I stopped learning mathematics and started studying how people think about it—my PhD is in mathematics education. Well, I say I stopped learning, but of course I didn't. In studying how people think about something, you inevitably learn more about it. I've also taught a lot of undergraduate mathematics, and there's nothing like teaching something to make you understand it properly.

But, while I teach mathematics, I don't create it—I don't spend my working life developing new mathematical ideas. That's what mathematicians do: they conduct research that extends our understanding of abstract mathematical structures and research that has real-world applications. The mathematicians down the corridor from me at Loughborough University do this in a variety of fields. Claudia Garetto studies hyperbolic equations, which is a subfield of partial differential equations with connections to physics; her methods combine geometry with analysis, involving rigorous study of conditions under which certain types of equation have unique solutions. Natalia Janson studies spontaneously evolving systems with complex behaviour, including plasma physics, electronic circuits, and the cardiovascular and nervous systems; her work can predict the emergence of chaos or synchronization in man-made devices or living systems. Eugénie Hunsicker researches the statistical analysis of images, which improves our ability to monitor production of the high-specification materials used in computing, electronics, and the space and nuclear industries. Diwei Zhou conducts research on non-Euclidean statistics, which can be applied in diffusion magnetic resonance imaging (MRI) to improve our ability to visualize and analyse the microstructure of biological tissue. Like all academics, these mathematicians communicate via conferences and academic journals. If you want to impress yourself with some cutting edge mathematical terminology—and probably learn new symbols while you're at it—you could look up articles in journals like *Annals of Mathematics*, *Acta Mathematica*, the *SIAM Journal on Applied Mathematics*, and *Physical Review Letters* (there is a lot of crossover at the border of applied mathematics and physics).

Researchers like me, in contrast, develop new knowledge about mathematical thinking. The Mathematics Education Centre, where I work, houses people who studied mathematics then shifted into education, and people who studied psychology then focused on mathematical cognition. Some of us study mathematical thinking for its own sake, and some use mathematical thinking to study general phenomena: arithmetic provides an unusually tidy set of logically related things, so it is useful for studying working memory, anxiety, inhibition, and so on. I study students' learning at the transition to university, focusing on their logical reasoning and understanding of proofs; I recently worked on the eye-movement research discussed in the Introduction. Nina Attridge investigates the effects of physical pain on numerical thinking; her research shows that people in pain may make poorer numerical and financial decisions, and she aims to develop interventions to help people living with chronic pain to overcome this effect. Sophie Batchelor studies skills and dispositions that help children in early stages of number learning; she is currently running an international study with colleagues in Northern Ireland, Belgium, and Finland to investigate the effects of formal number instruction on children's early numerical skills. Camilla Gilmore investigates how individuals represent and process numbers and mathematical ideas, which includes uncovering how young children first come to understand the meaning of numbers; this work can help to reveal why some individuals have particular problems in learning mathematics and how we might help to overcome this. Iro Xenidou-Dervou conducts research on cognitive factors that influence children's early mathematics learning and achievement, such as working memory, IQ, language, magnitude processing, estimation, and counting skills; she is also examining how language affects adults' numerical cognition. To see the cutting edge of this kind of work, you could look at academic journals such as the *Journal for Research in Mathematics Education*, *Educational Studies in Mathematics*, or the *Journal of Numerical Cognition*, or at articles on mathematics learning in more generalist journals such as *Cognition* or *Learning and Instruction*.

What shall I read next?

I'm not about to suggest that you jump into reading journal articles, though. Mathematics is brilliant but it's unforgivingly hierarchical—if you haven't already done a degree, it could take several years of full-time study to understand even the titles. And articles in mathematics education aren't much easier—academics write for other academics, so they tend to assume knowledge of existing theories and methodological approaches. Fortunately, though, there are numerous accessible books for the mathematically interested. I hesitate to provide specific suggestions, because I haven't read everything and I wouldn't want to restrict anyone. If you want to learn more, I'd recommend that you spend some time browsing and see what catches your interest. But here are some comments that you might find useful.

Some popular mathematics books are more intellectual and some are more entertaining. Some present curious nuggets of mathematics, and some give more extended arguments. Some are about the historical development of mathematical ideas, or about their applications. And some authors have written on several of these topics across different books. Contemporary authors you might want to investigate include David Acheson, Alex Bellos, Keith Devlin, Marcus du Sautoy, Rob Eastaway, Martin Gardner, Timothy Gowers, Peter Higgins, Matt Parker, Chris Sangwin, Simon Singh, Ian Stewart, and Steven Strogatz. Some of these authors are prominent professional mathematicians, so you can learn about their research, too, via simple internet searches.

If you would like more experience of *doing* mathematics—maybe you enjoyed following this book's logical reasoning and you want to try some of that in a guided but more independent way—then a good place to start would be John Mason, Leone Burton, and Kaye Stacey's book *Thinking Mathematically*. If you want to pursue similar ideas at a more sophisticated level, I think R. B. J. T. Allenby's *Numbers and Proofs* is very engaging. If you would like to see mathematical reasoning applied to numerous problems, try Heinrich Dorrie's *100 Great Problems of Elementary Mathematics: Their History and Solution*. If you want to read about mathematical problem solving in general, then George Polyá's *How to Solve It* is the classic, and his two-part book *Mathematics and Plausible*

Reasoning expands on creative reasoning based on experience, examples, and analogies.

If you want to consider mathematics in relation to education—perhaps you are a parent or carer or teacher—you might consider Rob Eastaway and Mike Askew's *Maths for Mums and Dads* books, Anne Watson, Keith Jones, and Dave Pratt's *Key Ideas in Teaching Mathematics*, or Colin Foster's books, including *Questions Pupils Ask*. Both are informed by research as well as by practical classroom experience. And there are many more books for teachers. Those that focus on teaching about links between mathematical concepts include Magdalene Lampert's *Teaching Problems and the Problems of Teaching* and Liping Ma's *Knowing and Teaching Elementary Mathematics*. If you're interested in a cognitive psychology perspective, try Camilla Gilmore, Silke Göbel, and Matthew Inglis' *An Introduction to Mathematical Cognition*. Or, to learn about the relationship between mathematics and logical reasoning, Matthew Inglis and Nina Attridge's *Does Mathematical Study Develop Logical Thinking?*

If you'd like a book similar to this one in style, I recommend W. W. Sawyer's classics *Mathematician's Delight* and *Prelude to Mathematics*, and Cornelius Lanczos's *Numbers Without End*, which is hard to get hold of but which inspired me as a teenager. If you'd like something similar in style but different in content, try Jordan Ellenberg's *How Not to Be Wrong: The Hidden Mathematics of Everyday Life*. Ellenberg's very entertaining book is about applying careful mathematical reasoning to everyday situations, particularly those that involve making decisions under uncertainty. For something more applied, Paul Lockhart's *Measurement* covers both size and shape and time and space, and retains its conceptual focus and informal style while making a speedier transition to advanced ideas. And if you're a teenager or a returner to education who is planning to study mathematics at university, I hope you'll consider my books *How to Study for a Mathematics Degree* (or its North American equivalent *How to Study as a Mathematics Major*) and *How to Think about Analysis*.

Finally, if you've had enough of mathematics for now but you've improved your knowledge, your confidence, or both, then I'm delighted. Thank you for reading.

REFERENCES

Acevedo Nistal, A., Van Dooren, W., & Verschaffel, L. (2013). Students' reported justifications for their representational choices in linear function problems: An interview study. *Educational Studies, 39,* 104–117.

Ainsworth, S. (2006). DeFT: A conceptual framework for considering learning with multiple representations. *Learning and Instruction, 16,* 183–198.

Ainsworth, S., & Burcham, S. (2007). The impact of text coherence on learning by self-explanation. *Learning and Instruction, 17,* 286–303.

Ainsworth, S., & Th Loizou, A. (2003). The effects of self-explanation when learning with text or diagrams. *Cognitive Science, 27,* 669–681.

Alcock, L. (2010). Mathematicians' perspectives on the teaching and learning of proof. In F. Hitt, D. Holton, & P. W. Thompson (Eds.), *Research in collegiate mathematics education VII* (pp. 63–92). Washington, DC: MAA.

Alcock, L. (2013a). *How to study as a mathematics major.* Oxford: Oxford University Press.

Alcock, L. (2013b). *How to study for a mathematics degree.* Oxford: Oxford University Press.

Alcock, L. (2014). *How to think about analysis.* Oxford: Oxford University Press.

Alcock, L., Bailey, T., Inglis, M., & Docherty, P. (2014). The ability to reject invalid logical inferences predicts proof comprehension and mathematics performance. In *Proceedings of the 17th conference on research in undergraduate mathematics education.* Denver, CO.

Alcock, L., Hodds, M., Roy, S., & Inglis, M. (2015). Investigating and improving undergraduate proof comprehension. *Notices of the American Mathematical Society, 62,* 742–752.

Alcock, L., & Inglis, M. (2008). Doctoral students' use of examples in evaluating and proving conjectures. *Educational Studies in Mathematics, 69,* 111–129.

Alcock, L., & Simpson, A. (2002). Definitions: dealing with categories mathematically. *For the Learning of Mathematics, 22*(2), 28–34.

Alcock, L., & Simpson, A. (2004). Convergence of sequences and series: Interactions between visual reasoning and the learner's beliefs about their own role. *Educational Studies in Mathematics, 57*, 1–32.

Alcock, L., & Simpson, A. (2017). Interactions between defining, explaining and classifying: the case of increasing and decreasing sequences. *Educational Studies in Mathematics, 94*, 5–19.

Alcock, L., & Weber, K. (2010). Referential and syntactic approaches to proving: Case studies from a transition-to-proof course. In F. Hitt, D. Holton, & P. W. Thompson (Eds.), *Research in collegiate mathematics education vii* (pp. 93–114). Washington, DC: MAA.

Alibali, M. W., & Sidney, P. G. (2015). Variability in the natural number bias: Who, when, how and why. *Learning and Instruction, 37*, 56–61.

Anghileri, J. (1989). An investigation of young children's understanding of multiplication. *Educational Studies in Mathematics, 20*, 367–385.

Ariasi, N., & Mason, L. (2011). Uncovering the effect of text structure in learning from a science text: An eye-tracking study. *Instructional Science, 39*, 581–601.

Attridge, N., Doritou, M., & Inglis, M. (2015). The development of reasoning skills during compulsory 16 to 18 mathematics education. *Research in Mathematics Education, 17*, 20–37.

Attridge, N., & Inglis, M. (2013). Advanced mathematical study and the development of conditional reasoning skills. *PLoS ONE, 8*, e69399.

Bell, A. W. (1976). A study of pupils' proof conceptions in mathematical situations. *Educational Studies in Mathematics, 7*, 23–40.

Bielaczyc, K., Pirolli, P. L., & Brown, A. L. (1995). Training in self-explanation and self-regulation strategies: Investigating the effects of knowledge acquisition activities on problem solving. *Cognition and Instruction, 13*, 221–252.

Bjork, R. A., Dunlosky, J., & Kornell, N. (2013). Self-regulated learning: Beliefs, techniques, and illusions. *Annual Review of Psychology, 64*, 417–444.

Booth, J. L., Lange, K. E., Koedinger, K. R., & Newton, K. J. (2013). Using example problems to improve student learning in algebra: Differentiating between correct and incorrect examples. *Learning and Instruction, 25*, 24–34.

Brown, J. R. (1999). *Philosophy of mathematics: An introduction to the world of proofs and pictures*. New York: Routledge.

Buchbinder, O., & Zaslavsky, O. (2011). Is this a coincidence? The role of examples in fostering a need for proof. *ZDM: International Journal on Mathematics Education, 43*, 269–281.

Burn, R. P. (1992). *Numbers and functions: Steps into analysis*. Cambridge, UK: Cambridge University Press.

Cangelosi, R., Madrid, S., Cooper, S., Olson, J., & Hartter, B. (2013). The negative sign and exponential expressions: Unveiling students' persistent errors and misconceptions. *Journal of Mathematical Behavior, 32*, 69–82.

Castro Superfine, A., Canty, R. S., & Marshall, A. M. (2009). Translation between external representation systems in mathematics: All-or-none or skill conglomerate? *Journal of Mathematical Behavior, 28*, 217–236.

Chater, N., Heit, E., & Oaksford, M. (2005). Reasoning. In K. Lamberts & R. Goldstone (Eds.), *Handbook of cognition* (pp. 297–320). London: Sage.

Chazan, D. (1993). High school geometry students' justification for their views of empirical evidence and mathematical proof. *Educational Studies in Mathematics, 24*, 359–387.

Chi, M. T. H., Leeuw, N. D., Chiu, M.-H., & LaVancher, C. (1994). Eliciting selfexplanations improves understanding. *Cognitive Science, 18*, 439–477.

Chiu, M. M., & Klassen, R. M. (2010). Relations of mathematics self-concept and its calibration with mathematics achievement: Cultural differences among fifteen-year-olds in 34 countries. *Learning and Instruction, 20*, 2–17.

Copes, L. (1982). The perry development scheme: A metaphor for learning and teaching mathematics. *For the Learning of Mathematics, 3*(1), 38–44.

Cornu, B. (1991). Limits. In D. O. Tall (Ed.), *Advanced mathematical thinking* (pp. 153–166). Springer.

Cottrill, J., Dubinsky, E., Nichols, D., Schwingendorf, K., Thomas, K., & Vidakovic, D. (1996). Understanding the limit concept: Beginning with a coordinated process scheme. *Journal of Mathematical Behavior, 15*, 167–192.

Cowen, C. (1991). Teaching and testing mathematics reading. *American Mathematical Monthly, 98*, 50–53.

Crawford, K., Gordon, S., Nicholas, J., & Prosser, M. (1994). Conceptions of mathematics and how it is learned: The perspectives of students entering university. *Learning and Instruction, 4*, 331–345.

Cromley, J. G., Synder-Hogan, L. E., & Luciw-Dubas, U. A. (2010). Cognitive activities in complex science text and diagrams. *Contemporary Educational Psychology, 35,* 59–74.

Dahlberg, R. P., & Housman, D. L. (1997). Facilitating learning events through example generation. *Educational Studies in Mathematics, 33,* 283–299.

Davis, P., & Hersh, R. (1983). *The mathematical experience.* Harmondsworth: Penguin.

Davis, R. B., & Vinner, S. (1986). The notion of limit: Some seemingly unavoidable misconception stages. *Journal of Mathematical Behavior, 5,* 281–303.

De Bock, D., Verschaffel, L., Janssens, D., Van Dooren, W., & Claes, K. (2003). Do realistic contexts and graphical representations always have a beneficial impact on students' performance? Negative evidence from a study on modelling nonlinear geometry problems. *Learning and Instruction, 13,* 441–463.

de Villiers, M. (1990). The role and function of proof in mathematics. *Pythagoras, 24,* 17–24.

DeWolf, M., & Vosniadou, S. (2015). The representation of fraction magnitudes and the whole number bias reconsidered. *Learning and Instruction, 37,* 39–49.

Dreyfus, T. (1991). On the status of visual reasoning in mathematics and mathematics education. In F. Furinghetti (Ed.), *Proceedings of the 15th International Conference on the Psychology of Mathematics Education* (Vol. 1, pp. 33–48). Assissi, Italy: IGPME.

Dubinsky, E., Dautermann, J., Leron, U., & Zazkis, R. (1994). On learning fundamental concepts of group theory. *Educational Studies in Mathematics, 27,* 267–305.

Durkin, K., & Rittle-Johnson, B. (2012). The effectiveness of using incorrect examples to support learning about decimal magnitude. *Learning and Instruction, 22,* 206–214.

Durkin, K., & Rittle-Johnson, B. (2015). Diagnosing misconceptions: Revealing changing decimal fraction knowledge. *Learning and Instruction, 37,* 21–29.

Epp, S. (2003). The role of logic in teaching proof. *American Mathematical Monthly, 110,* 886–899.

Evans, J. S. B. T. (2007). *Hypothetical thinking: Dual processes in reasoning and judgement.* Hove, UK: Psychology Press.

Evans, J. S. B. T., & Over, D. E. (1996). *Rationality and reasoning*. Hove, UK: Psychology Press.

Evans, J. S. B. T., & Over, D. E. (2004). *If*. Oxford: Oxford University Press.

Fischbein, E. (1982). Intuition and proof. *For the Learning of Mathematics, 3*(2), 9–18.

Fujita, T., & Jones, K. (2007). Learners' understanding of the definitions and hierarchical classification of quadrilaterals: Towards a theoretical framing. *Research in Mathematics Education, 9*, 3–20.

Fyfe, E. R., McNeil, N. M., & Borjas, S. (2015). Benefits of 'concreteness fading' for children's mathematics understanding. *Learning and Instruction, 35*, 104–120.

Gagatsis, A., & Shiakalli, M. (2004). Ability to translate from one representation of the concept of function to another and mathematical problem solving. *Educational Psychology, 24*, 645–657.

Giaquinto, M. (2007). *Visual thinking in mathematics*. Oxford: Oxford University Press.

Gilmore, C., & Spelke, E. S. (2008). Children's understanding of the relationship between addition and subtraction. *Cognition, 107*, 932–945.

Gilmore, C. K., McCarthy, S., & Spelke, E. S. (2007). Symbolic arithmetic knowledge without instruction. *Nature, 447*, 589–591.

Goldenberg, P., & Mason, J. (2008). Shedding light on and with example spaces. *Educational Studies in Mathematics, 69*, 183–194.

Goldin, G., & Shteingold, N. (2001). Systems of representations and the development of mathematical concepts. In A. A. Cuoco & F. R. Curcio (Eds.), *The roles of representation in school mathematics* (pp. 1–23). Reston, VA: National Council of Teachers of Mathematics.

Gray, E., & Tall, D. (1994). Duality, ambiguity and flexibility: A proceptual view of simple arithmetic. *Journal for Research in Mathematics Education, 25*, 115–141.

Hadamard, J. (1945). *The psychology of invention in the mathematical field* (1954 ed.). New York: Dover.

Hanna, G. (1991). Mathematical proof. In D. O. Tall (Ed.), *Advanced mathematical thinking* (pp. 54–61). Dordrecht: Kluwer.

Harel, G. (2001). The development of mathematical induction as a proof scheme: A model for DNR-based instruction. In S. Campbell & R. Zazkis (Eds.), *Learning and teaching number theory* (pp. 185–212). Westport, CT: Ablex.

Harel, G., & Sowder, L. (1998). Students' proof schemes: Results from exploratory studies. In A. H. Schoenfeld, J. Kaput, & E. Dubinsky (Eds.), *Research in collegiate mathematics III* (pp. 234–282). Providence, RI: American Mathematical Society.

Harel, G., & Tall, D. (1989). The general, the abstract, and the generic in advanced mathematics. *For the Learning of Mathematics, 11* (1), 38–42.

Healy, L., & Hoyles, C. (2000). A study of proof conceptions in algebra. *Journal for Research in Mathematics Education, 31*, 396–428.

Heinze, A. (2010). Mathematicians' individual criteria for accepting theorems and proofs: An empirical approach. In G. Hanna, H. N. Jahnke, & H. Pulte (Eds.), *Explanation and proof in mathematics* (pp. 101–111). New York: Springer.

Hersh, R. (1993). Proving is convincing and explaining. *Educational Studies in Mathematics, 24*, 389–399.

Hoch, M., & Dreyfus, T. (2004). Structure sense in high school algebra: The effect of brackets. In M. J. Hoines & A. B. Fuglestad (Eds.), *Proceedings of the 28th conference of the international group for the psychology of mathematics education* (Vol. 3, pp. 49–56). Bergen, Norway: IGPME.

Hodds, M., Alcock, L., & Inglis, M. (2014). Self-explanation training improves proof comprehension. *Journal for Research in Mathematics Education, 45*, 62–101.

Hoyles, C., & Küchemann, D. (2002). Students' understanding of logical implication. *Educational Studies in Mathematics, 51*, 193–223.

Iannone, P., Inglis, M., Mejía-Ramos, J., Simpson, A., & Weber, K. (2011). Does generating examples aid proof production? *Educational Studies in Mathematics, 77*, 1–14.

Inglis, M., & Alcock, L. (2012). Expert and novice approaches to reading mathematical proofs. *Journal for Research in Mathematics Education, 43*, 358–390.

Inglis, M., & Mejía-Ramos, J. P. (2009). On the persuasiveness of visual arguments in mathematics. *Foundations of Science, 14*, 97–110.

Inglis, M., Mejía-Ramos, J.-P., Weber, K., & Alcock, L. (2013). On mathematicians' different standards when evaluating elementary proofs. *Topics in Cognitive Science, 5*, 270–282.

Inglis, M., & Simpson, A. (2008). Conditional inference and advanced mathematical study. *Educational Studies in Mathematics, 67*, 187–204.

Inglis, M., & Simpson, A. (2009). Conditional inference and advanced mathematical study: Further evidence. *Educational Studies in Mathematics, 72*, 185–198.

Jansen, A. R., Marriott, K., & Yelland, G. W. (2003). Comprehension of algebraic expressions by experienced users of mathematics. *Quarterly Journal of Experimental Psychology, 56A*, 3–30.

Johnson-Laird, P. N., & Byrne, R. M. J. (1991). *Deduction*. Hove, UK: Erlbaum.

Jones, M. G. (2009). Transfer, abstraction, and context. *Journal for Research in Mathematics Education, 40*, 80–89.

Kirshner, D., & Awtry, T. (2004). Visual salience of algebraic transformations. *Journal for Research in Mathematics Education, 35*, 224–257.

Knuth, E. J., Stephens, A. C., McNeil, N. M., & Alibali, M. W. (2006). Does understanding the equal sign matter? evidence from solving equations. *Journal for Research in Mathematics Education, 37*, 297–312.

Küchemann, D. (1981). Algebra. In K. Hart (Ed.), *Children's understanding of mathematics: 11–16* (pp. 102–119). Oxford: Alden Press.

Lai, Y., Weber, K., & Mejía-Ramos, J.-P. (2012). Mathematicians' perspectives on features of a good pedagogical proof. *Cognition and Instruction, 30*, 146–169.

Lakatos, I. (1976). *Proofs and refutations*. Cambridge, UK: Cambridge University Press.

Landy, D., & Goldstone, R. (2007a). How abstract is symbolic thought? *Journal of Experimental Psychology: Learning, Memory and Cognition, 33*, 720–733.

Landy, D., & Goldstone, R. L. (2007b). Formal notations are diagrams: Evidence from a production task. *Memory & Cognition, 35*, 2033–2040.

Larsen, S. P. (2013). A local instructional theory for the guided reinvention of the group and isomorphism concepts. *Journal of Mathematical Behavior, 32*, 712–725.

Leinhardt, G., Zaslavsky, O., & Stein, M. K. (1990). Functions, graphs, and graphing: Task, learning, and teaching. *Review of Educational Research, 60*, 1–64.

Leron, U. (1985). A direct approach to indirect proofs. *Educational Studies in Mathematics, 16*, 321–325.

Levenson, E. (2012). Teachers' knowledge of the nature of definitions: The case of the zero exponent. *Journal of Mathematical Behavior, 31*, 209–219.

Levenson, E., Tirosh, D., & Tsamir, P. (2009). Students' perceived socio-mathematical norms: The missing paradigm. *Journal of Mathematical Behavior, 28*, 171–187.

Lithner, J. (2003). Students' mathematical reasoning in university textbook exercises. *Educational Studies in Mathematics, 52*, 29–55.

Lowrie, T., & Kay, R. (2001). Relationship between visual and nonvisual solution methods and difficulty in elementary mathematics. *Journal of Educational Research, 94*, 248–255.

McNamara, D. S., Kintsch, E., Songer, N. B., & Kintsch, W. (1996). Are good texts always better? Interactions of text coherence, background knowledge, and levels of understanding in learning from text. *Cognition and Instruction, 14*, 1–43.

Mason, J. (2000). Asking mathematical questions mathematically. *International Journal of Mathematical Education in Science and Technology, 31*, 97–111.

Mason, J., & Pimm, D. (1984). Generic examples: Seeing the general in the particular. *Educational Studies in Mathematics, 15*, 277–289.

Mejía-Ramos, J.-P., & Weber, K. (2014). Why and how mathematicians read proofs: Further evidence from a survey study. *Educational Studies in Mathematics, 85*, 161–173.

Michener, E. R. (1978). Understanding understanding mathematics. *Cognitive Science, 2*, 361–383.

Mills, M. (2014). A framework for example usage in proof presentations. *Journal of Mathematical Behavior, 33*, 106–118.

Oaksford, M., & Chater, N. (2007). *Bayesian rationality: The probabilistic approach to human reasoning.* Oxford: Oxford University Press.

Obersteiner, A., & Tumpek, C. (2016). Measuring fraction comparison strategies with eye-tracking. *ZDM Mathematics Education, 48*, 255–266.

Obersteiner, A., Van Dooren, W., Van Hoof, J., & Verschaffel, L. (2013). The natural number bias and magnitude representation in fraction comparison by expert mathematicians. *Learning and Instruction, 28*, 64–72.

Österholm, M. (2005). Characterizing reading comprehension of mathematical texts. *Educational Studies in Mathematics, 63*, 325–346.

Peled, I., & Zaslavsky, O. (1997). Counter-examples that (only) prove and counterexamples that (also) explain. *Focus on Learning Problems in Mathematics, 19*, 49–61.

Poincaré, H. (1905). *Science and hypothesis.* London: Walter Scott.

Presmeg, N. C. (1986). Visualsation and mathematical giftedness. *Educational Studies in Mathematics, 17,* 297–311.

Raman, M. (2003). Key ideas: What are they and how can they help us understand how people view proof? *Educational Studies in Mathematics, 52,* 319–325.

Renkl, A. (2002). Worked-out examples: Instructional explanations support learning by self-explanations. *Learning and Instruction, 12,* 529–556.

Rowland, T. (2002). Generic proofs in number theory. In S. R. Cambpell & R. Zazkis (Eds.), *Learning and teaching number theory: Research in cognition and instruction* (pp. 157–184). Westport, CT: Ablex.

Rowland, T. (2008). The purpose, design and use of examples in the teaching of elementary mathematics. *Educational Studies in Mathematics, 69,* 149–163.

Schoenfeld, A. H. (1985). *Mathematical problem solving.* San Diego: Academic Press.

Schoenfeld, A. H. (1992). Learning to think mathematically: problem solving, metacognition and sense making in mathematics. In D. Grouws (Ed.), *Handbook of research on mathematics teaching and learning* (pp. 334–370). New York: Macmillan.

Selden, A., & Selden, J. (2003). Validations of proofs considered as texts: can undergraduates tell whether an argument proves a theorem? *Journal for Research in Mathematics Education, 34,* 4–36.

Selden, J., & Selden, A. (1995). Unpacking the logic of mathematical statements. *Educational Studies in Mathematics, 29,* 123–151.

Sfard, A. (1991). On the dual nature of mathematical conceptions: Reflections on processes and objects as different sides of the same coin. *Educational Studies in Mathematics, 22,* 1–36.

Shepherd, M. D., & van de Sande, C. C. (2014). Reading mathematics for understanding—from novice to expert. *Journal of Mathematical Behavior, 35,* 74–86.

Siegler, R. S., & Lortie-Forgues, H. (2015). Conceptual knowledge of fraction arithmetic. *Journal of Educational Psychology, 107,* 909–918.

Skemp, R. R. (1976). Relational understanding and instrumental understanding. *Mathematics Teaching, 77,* 20–26.

Stanovich, K. E., & West, R. F. (1998). Individual differences in rational thought. *Journal of Experimental Psychology: General, 127,* 161–188.

Stylianou, D. A., & Silver, E. A. (2004). The role of visual representations in advanced mathematical problem solving: An examination of expert-novice similarities and differences. *Mathematical Thinking and Learning, 6*, 353–387.

Tall, D. (2013). *How humans learn to think mathematically.* Cambridge, UK: Cambridge University Press.

Tall, D. O. (1995). Cognitive development, representations and proof. In *Proceedings of justifying and proving in school mathematics* (pp. 27–38). London: IoE.

Tall, D. O., & Vinner, S. (1981). Concept image and concept definition in mathematics with particular reference to limits and continuity. *Educational Studies in Mathematics, 12*, 151–169.

Thurston, W. P. (1994). On proof and progress in mathematics. *Bulletin of the American Mathematical Society, 30*, 161–177.

Torbeyns, J., Schneider, M., Xin, Z., & Siegler, R. S. (2015). Bridging the gap: Fraction understanding is central to mathematics achievement in students from three different continents. *Learning and Instruction, 37*, 5–13.

Tsamir, P. (2003). Using the intuitive rule more A—more B for predicting and analysing students' solutions in geometry. *International Journal of Mathematical Education in Science and Technology, 34* 639–650.

Tsamir, P., Tirosh, D., & Levenson, E. (2008). Intuitive nonexamples: The case of triangles. *Educational Studies in Mathematics, 49*, 81–95.

Vamvakoussi, X., Christou, K. P., Mertens, L., & Van Dooren, W. (2011). What fills the gap between discrete and dense? Greek and Flemish students' understanding of density. *Learning and Instruction, 21*, 676–685.

Vamvakoussi, X., & Vosniadou, S. (2004). Understanding the structure of the set of rational numbers: A conceptual change approach. *Learning and Instruction, 14*, 453–467.

Vamvakoussi, X., & Vosniadou, S. (2010). How many decimals are there between two fractions? Aspects of secondary school students' understanding of rational numbers and their notation. *Cognition and Instruction, 28*, 181–209.

Van Dooren, W., de Bock, D., Weyers, D., & Verschaffel, L. (2004). The predictive power of intuitive rules: A critical analysis of 'more A—more B' and 'same A—same B'. *Educational Studies in Mathematics, 56*, 179–207.

Van Hoof, J., Vandewalle, J., Verschaffel, L., & Van Dooren, W. (2015). In search for the natural number bias in secondary school students' interpretation of the effect of arithmetical operations. *Learning and Instruction, 37*, 30–38.

Van Dooren, W., De Bock, D., Evers, M., & Verschaffel, L. (2009). Students' overuse of proportionality on missing-value problems: How numbers may change solutions. *Journal for Research in Mathematics Education, 40*, 187–211.

Vinner, S. (1991). The role of definitions in teaching and learning. In D. O. Tall (Ed.), *Advanced mathematical thinking* (pp. 65–81). Dordrecht: Kluwer.

Weber, K. (2008). How mathematicians determine if an argument is a valid proof. *Journal for Research in Mathematics Education, 39*, 431–459.

Weber, K. (2010). Proofs that develop insight. *For the Learning of Mathematics, 30*, 32–36.

Weber, K., & Alcock, L. (2004). Semantic and syntactic proof productions. *Educational Studies in Mathematics, 56*, 209–234.

Weber, K., Inglis, M., & Mejía-Ramos, J. P. (2014). How mathematicians obtain conviction: Implications for mathematics instruction and research on epistemic cognition. *Educational Psychologist, 49*, 36–58.

Wilkerson-Jerde, M. H., & Wilensky, U. J. (2011). How do mathematicians learn math?: Resources and acts for constructing and understanding mathematics. *Educational Studies in Mathematics, 78*, 21–43.

Winicki-Landman, G., & Leikin, R. (2000). On equivalent and non-equivalent definitions: Part 1. *For the Learning of Mathematics, 20*(1), 17–21.

Yackel, E., & Cobb, P. (1996). Sociomathematical norms, argumentation, and autonomy in mathematics. *Journal for Research in Mathematics Education, 27*, 458–477.

Zazkis, R., & Chernoff, E. J. (2008). What makes a counterexample exemplary? *Educational Studies in Mathematics, 68*, 195–208.

Zazkis, R., & Leikin, R. (2007). Generating examples: From pedagogical tool to a reserach tool. *For the Learning of Mathematics, 27*(2), 15–21.

Zazkis, R., & Leikin, R. (2008). Exemplifying definitions: A case of a square. *Educational Studies in Mathematics, 69*, 131–148.

Zazkis, R., Liljedahl, P., & Chernoff, E. J. (2008). The role of examples in forming and refuting generalizations. *ZDM: International Journal on Mathematics Education, 40*, 131–141.

Zhen, B., Weber, K., & Mejía-Ramos, J.-P. (2016). Mathematics majors' perceptions of the admissibility of graphical inferences in proofs. *International Journal of Research in Undergraduate Mathematics Education, 2*, 1–29.

INDEX

$0.3\dot{6}$ 84, 168
$0.\overline{36}$ 84, 168
$180°$ 51, 53, 54, 63, 152
$\frac{1}{2} \times \frac{1}{2}$ 11
$\frac{1}{2}bh$ 21, 23, 24
2^0 194
$2n - 1$ 98
$\frac{2}{7}$ 167, 176–8
$(3, 4, 5)$ triangle 30
$360°$ 48, 51, 53, 54, 59, 152
$\sqrt{3}$ 186, 187
$\frac{4}{11}$ 167, 168
$(a + b)^2$ 16, 17
$(a + b)^3$ 18
$(a + b + c)^2$ 17
$a^2 + b^2 = c^2$ 37
$a^3 + b^3 = c^3$ 37
$a^n + b^n = c^n$ 37
$\forall$ 127
$\Leftrightarrow$ 131
$\Rightarrow$ 54
$\approx$ 27, 184
$\equiv$ 19
$\neq$ 30
$\pm$ 150
$\ldots$ 81, 93
∞ 128, 129, 188, 189
$\aleph_0$ 189, 191, 192, 194
ϕ 156
π 183–6
θ 151, 155
$x^2 + y^2 = r^2$ 148
$x^2 - y^2$ 19, 20
$x^m x^n = x^{m+n}$ 102, 194, 195

$y = mx$ 126–9
$y = mx + c$ 130, 132

A
abstract 71, 75, 76, 80, 160
abstract algebra 76
accessibility 101
addend 82, 162
addition 76
   adding 9 160
   binary operation 6
   commutative 5, 6
   fractions 88, 89, 91, 115
   number line 196
   odd numbers 96, 97
advanced xv, xvi
aleph nought 189, 191, 192, 194
algebra 66
   $(a + b)^2$ 16, 17
   $(a + b)^3$ 18
   $(a + b + c)^2$ 17
   $x^2 - y^2$ 19, 20
   $x^m x^n = x^{m+n}$ 102, 194, 195
   arguments 96, 101
   consistency 194
   diagrams 16, 39
   difference of two squares 33
   fallacy 17, 39, 102, 118
   formulas 55
   general arguments 80
   intersecting lines 138, 139
   manipulation 66
   squaring 16, 17
aligned 42, 47
all 127

always 56, 80, 101
ambiguity 93, 108, 165
angles
  along a straight line 51, 53
  around a point 48, 53
  circle 53
  cylindrical polar coordinates 155
  formula 55
  general polygon 54
  interior 42, 43, 48, 49, 51–3, 55, 57, 63
  octagon 55
  pentagon 54
  polar coordinates 151
  spherical polar coordinates 156
  supplementary 51
  triangle 53, 54
annulus 153
answer 168
aperiodic 79
application 56, 80, 124, 205, 207, 208
approximation 168, 183, 184
area 8, 10, 11, 15, 21
  quadrilaterals 25
  rectangles 142–4
  squares 15
  triangles 21–4
argument 29, 201
  $\sqrt{3}$ 187
  algebraic 96, 101
  concise 100
  decimal to fraction 174, 175
  deductive 46, 96
  diagonalization 193
  divisible by 9 162, 163
  general 119, 162
  geometric series 108, 114
  harmonic series 112
  inductive 98–100, 104, 116, 119
  logical 97
  nonexistence 37, 63
  numerical 59
  visual 96, 101, 119, 190
arithmetic 9
  broken by infinity 129, 188

consistency 194–6
  infinite set 189
  negative numbers 196, 198
  with symmetries 75
array 2, 4–6, 190
associativity 76, 183
atypical 23, 39
audience 93, 101, 201
axioms 75, 76
axis
  crossing 124, 150, 154
  graph 122
  horizontal 122, 125
  labelling 136
  reflection symmetry 73
  vertical 122, 128, 129
  z 154, 155

B
base
  10 160, 163, 165, 198
  triangle 21, 23–5
basement 196
basic vi, xiii–v
beauty 68
believing it always works 101
binary operation 6, 9, 75
  commutative 6
brackets 6, 7
building blocks 70, 71

C
cakes 14, 141
calculation xiv, 199, 201
  adding fractions 89
  decimal expansion 173
  division 168
  interior angles 52
  long division 168, 170
  long multiplication 181
  mental 94
  multiplying fractions 12
  notation 89

putting off 66
repetitive 80
calculator 106, 168, 172, 178
Cameron 141
can't xv
cancellation 105, 119
Cantor 192
cardinal arithmetic 190
cardinality 189–92, 194
carers xiv, 208
carpentry 121, 135
carrots 56, 89
Cartesian coordinates 151–4, 158
categorization 45, 46
centre 150
chain
  of claims 119
  of equations 85
  of reasoning 101
change sides, change signs 68
children xiv, 166
circles
  angle in 53
  centre 150
  equation 147–9, 151, 158
  fraction representation 14, 15, 83
  polar equation 153
  to draw triangles 30
clarity 101
classification 79
closed 75
coefficients 131
cognition v, 206, 208
cognitive obstacle 10
common
  denominator 67, 90, 91
  factor 174, 187
  multiple 92
  ratio 105, 106, 114, 115, 119
communication 158, 167
commutativity 5, 6, 10, 16, 39, 94, 162, 183
comparison 83–5, 115, 118
compasses 30
components 70

composition 75, 76
compression 112
computer 32, 33, 45, 204
concise 100, 104
cone 157
confidence xiii, xiv, 113, 202, 208
confusion xvi
consistency 103, 189, 192, 194–9
constant 131
constraints 122, 135, 136
context 142, 167, 202
continuum 88, 185
contradiction 186, 187, 192, 193, 199
conventions 7, 45, 46, 158
  $y = mx + c$ 131
  graphing 124, 125
  polar coordinates 152
converge 114
converse 176
converting decimals to fractions 173–5,
  178
convex 46
coordinates 122, 123
  Cartesian 151–4, 158
  cylindrical polar 155
  polar 151–3, 155, 158
  spherical polar 156, 157
  three dimensions 154
corner 42
correspondence 189–2
countable 190–2, 194
countably infinite 190
counterintuitive 111, 113, 114, 150
counting 87, 108, 196
  finite sets 189
  infinite sets 189, 191
  numbers 189
  on 108
cross-multiplying 91
cube 18, 77
cubing 37
cuboid 155
curiosity 81
curriculum xv

curved graph 144, 146, 147
cylinders 155
cylindrical polar coordinates 155

D
Dalek 156
dart 78, 79
daunting xvi
decimal
  between 0 and 1 192
  comparison 84
  division 164
  expansion 84, 168, 198
  finite 168
  fractions 84, 173, 176, 178
  nonrepeating 185
  places 65
  repeating 84, 168, 171, 172, 174, 176–8,
    184, 185, 198, 199
  representation 164
  terminating 168, 174, 176
decision 46, 53, 122, 125, 195
deductive 46, 201
deer in the headlights 121
defining 45, 46, 79, 194, 195
definitions 46, 79
degenerate 39
  adding one thing 98
  in relation to typical 35
  multiplying by 1 9
  Pythagorean triple 34
  terminating decimals 176
  triangle 34, 35
denominator 12, 67, 83, 86, 89, 90, 174,
  179
  common 67, 90, 91
Descartes 151
diagonal 70
diagonalization argument 193
diagrams 20, 201, 202
  adding odd numbers 96, 97
  geometric series 109, 110
  information all at once 101
  insight 21

limitations 20, 21, 39, 96
Pythagoras' theorem 26, 29
semi-regular tessellations 59
sum 1 to 7 95
Venn 166, 176
diamond 44
difference of two squares 19, 20, 33
digit 163
dimensions
  three 18, 154, 155
  two 151
discrete 87, 88, 185
distributivity 6, 7, 10, 39, 94, 162, 197, 198
diverge 114
division
  by 3 164, 179
  by 5 165
  by 9 160–3, 167, 198
  decimals 164
  fractions 85, 167
  long 168–72, 176, 177, 179, 199
divisors 48, 53, 164
dodecagons 43, 62
domain 147
door 81, 107
doubling 105
Douglas 121, 135
drop a perpendicular 22, 148

E
edge 42
elegance xv, 105
ellipse 150
ellipsis 81, 93, 107, 113, 118, 168
enclosure 141, 144
epiphany xiv
equations 19, 39
  chain 85
  circle 147–9, 151, 158
  doing same to both sides 68
  for values of n 98
  graph 123, 124, 133

identities 19, 39
line 133, 151, 158
linear 127
multiplying by a constant 139
plane in three dimensions 154
polar circle 153
quadratic 146, 147
reading 67
rearranging 131, 133
related 98
simplifying 66
simultaneous 139
solving 66–8
equilateral triangle 41, 43, 48, 50
equivalent
⇔ 131
fractions 84, 118, 174, 178
estimate 82
even numbers 189
evidence 34, 53, 201, 202
example 37, 100, 118, 202
exercises xiii, xv, 26
expansion 84
experiment 74
expert xv, xviii, xix, 38, 149
explanation 195
exploring xvi, 51, 62, 92
expression 133
eye movements xvii, xix
eye tracker xviii

F
fact 195
factors 53, 67, 179, 180, 182, 183
common 174, 187
prime 71, 180, 182, 183, 186, 187
tree 179, 181
failure xiv
false 165
farmer 141, 142, 144
feasible 136, 137
fence 141, 144
Fermat's Last Theorem 1, 36–8

finite
decimal 168
sum 93, 107, 108
fixations xviii
fixed 38
flexible 189, 201
flip 73
food 45
for all 127
forever 81, 84, 107, 172
formulas
algebra 55
different 55
interior angles 52, 55, 66
labour-saving 52, 80
memorizing 25
Pythagoras' theorem 26
sum 1 to n 96
sum of geometric series 114
sum of odd numbers 97
sum of powers of 2 104
triangle area 24
understanding 26
using 26
foundational vi
fractions 11, 13, 14, 83
addition 88, 89, 91, 115
between 0 and 1 86
circles 14, 15, 83
comparison 85, 115
decimals 173, 176, 178
division 85, 167
equivalent 84, 118, 174, 178
greater or less 83
lowest terms 174
number line 86, 167
powers 195
proper 13, 175
rational numbers 176
ratios 84, 85, 118, 167
representation 85, 118, 183
sizes 88
splitting 90

fundamental 70
fussy 167

G
Gauss 93
generalization 9, 23, 37, 52, 56, 80, 94–6,
    119, 128, 160, 195
generate 69–71, 75
generic 5, 6, 10, 95, 96, 119, 162, 176
genius 94, 119
GeoGebra 64, 79
geometric series 107–10, 113–5, 119
geometry 30
-gon 43
gradient 126, 127, 132, 140
grammar 54
graph
    $y = mx$ 126
    curved 144, 146, 147
    equation 123, 124
    inequality 133, 135, 136
    negative numbers 124
    parabola 146
    problem solving 158
    profit 140
    straight line 132
Greek letter 151
group 75
group theory 75, 76, 80
guess 62

H
hard work 4
harmonic series 92, 110–4
Hazel 121, 135
hectagons 43
height 25
heptagons 61
heuristic 9–11, 39
hexagons 45, 50, 51, 60
    regular 42, 46, 48
hierarchical 207
history 38, 151, 195, 207

horizontal xv
    axis 122
    line 129, 137
hundreds, tens and units 160
hundredths 84, 172, 173
hypotenuse 26, 27, 148

I
ice cream 166, 167
identities 19, 39, 76
if 165, 167, 176, 199, 201
if and only if 29, 165, 167, 176, 178, 184,
    199
if…then… 99
imagination 85
implies 54, 201
independent 144
inductive argument 98–100, 104, 116
inequalities 133
    chain 85
    double 153
    graph 133, 135, 136
    regions 153, 155, 157, 158
infinite
    arithmetic 189
    countably 190
    subtraction 188
    sum 81, 82, 93, 110
    total for harmonic series 112
infinitesimal 109
infinity 128, 129
    breaks arithmetic 129, 188
    different infinities 188, 192
    intuition 129, 191
    is not a number 128, 188
    is really big 119
    multiplied by 0 128
insight xiv, xvii, 2, 5, 39
    diagrams 21, 119
    general algebra 99, 119
    lack of from calculator 168
    lack of from checking 98
    lack of from computer 33
    lack of from examples 119

integers 87, 118, 186, 187, 190
intercept 131, 132
interdependent 122
interesting difference 32, 37, 39, 42
interior angles 42, 48, 49, 51–5, 57, 63
introduction 1
intuition vi, 11, 59, 62, 109, 112, 119
    area and perimeter 142
    infinity 129, 190, 191
    optimization 144
    spherical polar coordinates 156
invariant 25, 39
    angles in pentagon 54
    number of dots 4, 6
    quadrilateral area 25
    triangle area 25
inverse 69, 71, 74, 76, 197
investigation
    generator symmetries 75
    Pythagorean triples 32, 36
    reflections 74
    semi-regular tessellations 42, 43, 58, 61
    symmetries 41, 68, 71
Iona 141
irrational numbers 183, 185–7, 192, 194,
    199

J
journals 205, 206
juxtaposition 7

K
kite 78, 79

L
labelling 125
labour saving 2, 52, 80
lazy 4, 179
levels 56
limit 108, 117
line 82, 159
    $x = 0$ 128
    $y = 0$ 127
    equation 151

graph 132, 133
    horizontal 129
    plotting 132
    profit 141
    vertical 129
linear equation 127
list 100, 190–3
logic xv, 165–7, 199, 201, 208
long division 168–72, 176, 177, 179, 199
lots of 11
Loughborough University xviii, 205
lowest terms 174, 175, 183, 187

M
magnitude 183
manipulation 66, 106
mathematical
    argument 29
    cognition v, 206, 208
    communication 100
    decision 46, 53, 125, 195
    difference 32, 37
    explanation 68
    grammar 54
    intuition 59
    language 201
    meaning 68, 80
    reading xiii, xvii, xviii, 54, 80, 149, 151,
        184
    relationships xiv, xviii, 103, 160
    sentences 54, 80
    structure 70
    surprise 58, 112, 113
    symbols 54
    theory 56, 80, 96
    thinking xv, 39, 206
mathematicians xiii, xvii, xviii, 4, 38, 40,
        83, 202, 205
mathematics
    as an enterprise 202
    creation 38, 151, 205
    evolving 38
    popular 207

mathematics (*continued*)
  pure 65, 157
  what is it for? 204
mathematics education v
Mathematics Education Centre xviii, 206
maximize 121, 135, 137, 147
meaning xiii, 68, 80, 89, 168, 194, 199
measurement 53
memorability xv
memorizing 3, 4, 25
mental arithmetic 106
milk 204
minus 149
mirror 68
mnemonics 68
multiplication
  $\frac{1}{2} \times \frac{1}{2}$ 11
  area 8, 11
  binary operation 6, 9
  by 10 173
  commutative 5, 6, 16
  facts 3
  fractions 11, 12
  grid 182
  juxtaposition 7, 16
  long 181
  lots of 11
  made easy 2
  makes things bigger 9, 10, 150
  negative numbers 197, 198
multivalued function 148

N
N-gon 63–5, 68
n-gon 52, 55
nths 56
naming 5
natural numbers 87, 189, 191, 192
necessary condition 166
negative numbers 21, 39, 124, 196
nervous xiv, 65, 112
nested 44, 45, 79
network xvi, 113

nonexistence argument 37, 63, 185
notation 7, 89, 102, 107, 108, 113, 163, 168, 201
number line 82, 86, 87, 89, 98, 118, 186, 192, 196
number systems 103, 159, 160, 194, 201
numbers
  ambiguity 93
  counting 87
  even 189
  group theory 75, 76
  integers 87, 118, 190
  irrational 183, 185–7, 192, 194, 199
  'next' rational 87, 190
  natural 87, 189, 191, 192
  negative 21, 39, 124, 196
  odd 96–8
  rational 13, 85, 87, 88, 118, 176, 178, 183–7, 190–2, 194, 199
  real 192, 194
  square 33, 35
  what are they? 159
  whole 87
numeral 159, 169
numerator 12, 83, 174, 179

O
obstacle 10
octagons 61
  regular 49
odd numbers 96–8
one-to-one correspondence 189–92
only if 165, 167, 176
optimization 121, 122, 135, 137–9, 141, 143–5, 147, 204
order xv
  listing integers 190
  listing rationals 190
  on number line 118
  reordering 94, 119
  shapes at a vertex 42
ordered pair 122, 152
origin 151

overdraft 196
overgeneralize 23

P
panic 121
parabola 146, 147
paradox 81
parallel 44, 145
parallelogram 44
parentheses 6
parents xiv, 208
pattern 201
pedantry 133
Penrose tiling 78, 79
pentagons 59
    regular 48, 52
perimeter 141–4
period 177, 178
perpendicular 22, 25, 44, 122, 125, 145, 148
persistent 189
pi 183–6
picture isn't a proof 96
pictures vi
pizzas 14
place value 160, 169, 173, 199
Platonist 195
polar coordinates 151–3, 155, 158
pole 151
polygons
    divide into triangles 54
    regular 47
powers
    fractional 195
    negative 102, 103, 164, 194, 195
    of 10 160, 164
    of 2 101–4
    zero 102, 103, 160, 194
practical 203
precision 31, 65, 84, 93, 167, 183, 201
prime factors 71, 180, 182, 183, 186, 187
problem solving 207
procedures xv
process 107, 108, 161, 168, 172, 177, 198
product 91

profit 121, 122, 135, 137–9, 141
proof 199
    by contradiction 186, 187, 192, 193, 199,
        201
    by induction 100, 104, 116, 119, 201
    concise 100, 101, 104, 119
    Fermat's Last Theorem 38
    picture isn't a 96
    structure 101
    sum odd numbers 100
    sum of powers of 2 104
    without words 96
proper
    fraction 13, 175
    subset 166
properties 201
psychology v, 206
pure mathematics 65, 157
Pythagoras' theorem 1, 26, 27, 29, 30,
        148, 151, 165
Pythagorean triple 31–6

Q
quadratic
    equation 146, 147
    function 146
quadrilaterals
    area 25
    nested categories 44, 45
    not rigid 44
qualifications xiv
questions xvi, 202

R
radius 148–50, 155
ratio
    change on graph 126
    common 105, 106, 114, 115, 119
    fractions 84, 85, 118
    term to predecessor 105
rational numbers 13, 85, 87, 88, 118, 176,
        178, 183–7, 190–2, 194, 199

reading xiii, xvii, xviii, 1, 80
  aloud 54, 80, 86, 149, 151, 184
  equations 67, 149, 151
real numbers 192, 194
real world 14, 124
rearranging 131, 133
reasoning
  linear 101
  long division 169
  rigorous 201
  valid 34
reciprocal 14
rectangles 10, 15
  area 142–4
  number of dots 96
  perimeter 142–4
rectangular prism 155
reflection 41, 68, 73–5, 77, 79
regions 153, 158
regular
  hexagon 46, 48
  octagon 49
  pentagon 48, 52
  polygon 42, 43, 47
  tessellation 47–9
relationships xiv, xviii, 4, 27, 103, 160
remainder 167, 168, 170, 176, 177
repeating
  decimal 84, 168, 171, 172, 174, 176–8,
    184, 185, 198, 199
  zeros 176, 177
repetitive xiii, xv, 80, 199
representations xvi, 4, 5, 201
  area 8, 10, 11, 15, 17, 21
  array 2, 4
  base 10 160, 163
  choice of 7
  circles for fractions 14, 15, 83
  decimals 84, 164
  diagrams 20, 21, 39
  dots 95
  fractions 85, 118
  graphs 158
  grid multiplication 182

insight 39
limitations 15, 20, 21, 39, 201
mental 85
numbers 183
translating between 158
visual vi, 95, 101, 143
rescaling 84
research xvii
  education xvi
  eye movements xvii–xix
  psychology v, xvi
  self-explanation training xix
restriction 45
right and wrong answers 125
right-angled triangle 21, 26–9, 148
rigid 43, 44, 201
room 81, 82
root 198
root 3 186, 187
rotation 41, 69, 71, 72, 77, 79
rounding 53, 65

S
saccades xviii
sandwiches 159
satisfaction 80
scanpath xvii, xviii
school xv, xvi, 38, 168
self-explanation training xix
semi-regular tessellation 42, 47, 49–51,
  58, 62, 68
sentences 54, 80
series 82, 111
  comparing 115, 118
  converge 114
  diverge 114
  geometric 107–10, 113–5
  harmonic 82, 92, 110–4
  shrinking terms 113, 114
  sum of $1/n^2$ 114, 116
simultaneous equations 139
slope 126
solids 155
solution set 133, 158

special case 15, 29, 47
specializing 33, 56
spheres 156, 157
spherical polar coordinates 156, 157
spreadsheet 32, 33
square number 33, 35
square roots 186, 198
squares 15, 42
    area 15
    as quadrilaterals 44
    difference of two 19, 20
    not rigid 44
    optimal solution 144
    optimization 145
    tessellation 47, 48
squaring 15, 17, 37, 186, 195
statements 165
steep 126
structure 70, 75, 76, 80, 101, 161, 165, 198,
    201, 205
struggle xiii
students xix, 121
subset 166
subtleties vi, 1, 8, 177
subtraction
    binary operation 6
    finite 188
    infinite 188
    not associative 76
    not commutative 6
    number line 196
sufficient condition 166
sum
    1 to 100 93, 94
    1 to 7 95
    1 to n 96
    finite 93, 107, 108
    for values of n 98
    geometric 105, 119
    infinite 81, 82, 93, 110
    odd numbers 96, 97, 100
    of digits 160–4, 179
    powers of 2 103, 104, 105
    visual representation 95

supermarkets 204
supplementary angle 51
surprise 58, 112, 113
swapping 2, 5
symbols 54, 80, 88, 93, 186
symmetrical 69, 73
symmetry 41, 68, 159, 201
    classification 77
    cube 77
    group theory 75, 76
    meaning 73
    reflection 41, 68, 73–5, 77, 79
    rotation 41, 69, 71, 72, 77, 79
    shapes 77
    transformation 73, 74
    translation 41, 69–71, 79
systematic 57, 179

T
tables 121, 122, 124, 135
talent xiv, 94
teachers xiii, xiv, 202, 203, 208
temperature 196
tennis xv
tenths 84, 171, 173
terminating decimal 168, 174, 176
terrible at maths xv
tessellation 41
    dodecagon 62
    heptagon 61
    hexagon 41, 60
    octagon 61
    pentagon 59
    regular 47–9
    semi-regular 42, 47, 49–51, 58, 62, 68
    square 41, 47, 48, 58
    triangle 41, 47, 58
    wibbly 58
tests xv, xvi
tetrahedron 155
Texas A&M University 79
textbooks 195

theorem 1, 2
  Fermat 1, 36–8
  if and only if 29
  Pythagoras 1, 26, 27, 29, 30, 148, 151, 165
theory vi, 56, 80, 96
three dimensions 18, 154, 155
tiling 41
  aperiodic 79
  Penrose 78, 79
time 107, 108, 122
times symbol 16
tracing paper 69, 72
transformation 73, 74
translation 41, 69–71, 79
  graph 130
triangles 22
  (3, 4, 5) 30
  area 21–4
  base 21, 23, 25
  circle equation 150
  drawing with circles 30
  equilateral 41, 43, 48, 50
  for circle equation 148
  height 25
  right-angled 21, 26–9, 148
  rigid 43
  tessellation 47
  total angles 53
  wrong way up 22, 23
triangular pyramid 155
true 165
turning 51–3, 56
typical 35

U
underbrace 102, 111
understanding xix, 26, 202, 203
unique 152, 182
units 160, 163

university 202
University of Oxford 79
University of Warwick 112

V
valid reasoning 34
variables 135
vegetables 166, 167
Venn diagrams 166, 176
vertex 42
vertical xvi
  axis 122, 128
  line 129, 137
  translation 130
vision xviii
visual
  argument 96, 101, 190
  representation vi, 95, 143

W
walking and turning 51–3, 56
wallpaper group 77
Warwick 112
weird 109, 113, 191
whole numbers 87
why 40, 63, 80, 99, 160, 168
word problem 121
words 80, 93

X
x-axis 122, 125, 127

Y
y-axis 122, 128, 129
y-intercept 131, 132

Z
z-axis 154, 155
zoom in 88